Postures for Non-Proliferation

Arms Limitation and Security Policies to Minimize Nuclear Proliferation

sipri

Stockholm International Peace Research Institute

SIPRI is an independent institute for research into problems of peace and conflict, especially those of disarmament and arms regulation. It was established in 1966 to commemorate Sweden's 150 years of unbroken peace.

The Institute is financed by the Swedish Parliament. The staff, the Governing Board and the Scientific Council are international. As a consultative body, the Scientific Council is not responsible for the views expressed in the publications of the Institute.

sipri

Stockholm International Peace Research Institute

Sveavägen 166, S-113 46 Stockholm, Sweden

Cable: Peaceresearch, Stockholm

Telephone: 08-15 09 40

Postures for Non-Proliferation

Arms Limitation and Security Policies to Minimize Nuclear Proliferation

Stockholm International Peace Research Institute

Taylor & Francis Ltd Crane, Russak & Company, Inc.
London
1979

First published 1979 by Taylor & Francis Ltd., London
and Crane, Russak & Company, Inc., New York

Library of Congress Cataloging in Publication Data

Stockholm International Peace Research Institute.
Postures for non-proliferation.

Written by E. C. B. Schoettle.
Bibliography: p.
1. Nuclear nonproliferation. 2. Arms control.
3. Military Policy. I. Schoettle, Enid C. B.
II. Title.
JX1974.73.S76 1979 327'.174 78-31210
ISBN 0-8448-1313-3

Typeset by Red Lion Setters, Holborn, London
Printed and bound in the United Kingdom by
Taylor & Francis (Printers) Ltd, Rankine Road,
Basingstoke, Hampshire RG24 0PR

Preface

The nature of the relationship between the arms limitation and security policies of the great powers — particularly the nuclear weapon states — and the acquisition of independent, nuclear weapon capabilities by additional non-nuclear weapon states is an important facet of the general problem of nuclear proliferation. However, it is not a question which has received much careful and systematic attention; rather, it is typically discussed in passing or in terms of particular arms limitation policies of the nuclear weapon states, on the one hand, or the incentives to 'go nuclear' of a particular non-nuclear weapon state, on the other. Yet systematic analysis of this relationship is warranted, since minimizing the proliferation of nuclear weapon capabilities is an important policy objective for world security and since the security policies of the major states help to determine the policy choices of other nations.

Thus, if one is interested in constraining nuclear proliferation, one must, in addition to designing a proliferation-resistant nuclear fuel cycle, identify a set of long-term arms limitation and security policies for the major states which are capable of satisfying various security and political objectives of non-nuclear weapon states, thereby minimizing their incentives to acquire independent nuclear weapon capabilities and collectively minimizing future nuclear proliferation. This book attempts to identify such a comprehensive arms limitation and security régime. It reviews negotiations concerning the NPT up to the end of the Review Conference of the NPT in mid-1975 and the strategic debate concerning nuclear proliferation.

Various books and monographs published by SIPRI over the past five years have addressed related aspects of nuclear non-proliferation. A list of these related SIPRI publications is appended to this text on page 161.

This book was written by Dr Enid C.B. Schoettle, who was a visiting scholar at SIPRI in the summers of 1974 and 1977. She was additionally supported by the Carnegie Endowment for International Peace; the Center for International Studies, Massachusetts Institute of Technology; and the Program for Science and International Affairs, Harvard University. The book was edited by Felicity Roos and Connie Wall. Dr Schoettle is now a Program Officer, Office of European and International Affairs, Ford Foundation.

January 1979

Frank Barnaby
Director

Contents

Preface

Chapter 1. Introduction 1
I. Definitions 1
II. Premises 5
The destabilizing effects of nuclear proliferation — Discrimination in the international arms limitation and security régime and in the NPT — Nuclear proliferation in a comprehensive and global perspective

Chapter 2. Arms limitation and security policies for minimizing nuclear proliferation: the strategic debate 14
I. Policy objectives of NNWS 14
Military security objectives — Political prestige objectives
II. Alternative strategies to satisfy policy objectives of NNWS 25
The High Posture Doctrine — The Low Posture Doctrine
III. Conclusion 66

Chapter 3. Negotiations on the NPT, 1965-68 74
I. Arms limitation and disarmament measures 77
The 1965 draft treaties — The 1967 identical draft treaties — The 1968 identical draft treaties — The March 1968 joint draft treaty — The May 1968 joint draft treaty
II. Security guarantees 99
The 1965 draft treaties — The 1967 identical draft treaties — The 1968 identical draft treaties — The 1968 tripartite draft Security Council resolution and March joint draft treaty — The May 1968 joint draft treaty
III. Provisions for review, duration and withdrawal 118
The 1965 draft treaties — The 1967 identical draft treaties — The 1968 identical draft treaties — The March and May 1968 joint draft treaties
IV. Conclusion 126

Chapter 4. The 1975 Review Conference of the Parties to the NPT 130
I. Arms limitation and disarmament measures 131
II. Security guarantees 138
III. Future review of the NPT 144

Chapter 5. Conclusion 146

References 148

Appendix 161

Index 163

Tables

Chapter 2. Arms limitation and security policies for minimizing nuclear proliferation: the strategic debate

2.1 The policy objectives of NNWS which might be achieved by acquiring a nuclear weapon capability 22
2.2 Effectiveness of alternative postures to satisfy various objectives of NNWS and thus minimize proliferation 72

1. Introduction

In order to design effective policies to minimize the proliferation of nuclear weapons, it is necessary to examine how the arms limitation and security policies of the major states in the international security system — particularly the nuclear weapon states — affect the decisions which non-nuclear weapon states must make concerning whether or not to acquire an independent nuclear weapon capability. The arms limitation and security policies which might minimize the future proliferation of nuclear weapons to additional states can perhaps best be identified, first, by analysing the debate concerning this relationship in the strategic literature on nuclear proliferation; and second, by analysing the discussions concerning this relationship which have gone on between the nuclear weapon states and the non-nuclear weapon states during negotiations on the Treaty on the Non-Proliferation of Nuclear Weapons (the NPT) of 1968 and at the Review Conference of the NPT held in 1975.

This introductory chapter will briefly define important terms and then discuss three premises on which the subsequent analysis is based. First, it is assumed that future proliferation of nuclear weapons to additional states would have destabilizing effects on the international security system. Second, it is assumed that the broader arms limitation and security régime in which the NPT is currently embedded constitutes a discriminatory international security régime. Finally, it is assumed that nuclear proliferation, and efforts to minimize it, must be analysed from a comprehensive and global perspective.

I. Definitions

Any analysis of the proliferation of nuclear weapons rests upon distinguishing classes of nation-states as a function of their current possession of, their potential for acquiring, or their lack of potential for acquiring an independent nuclear weapon capability.

The class of nuclear weapon states (NWS) is defined as those states which have "manufactured and exploded a nuclear weapon or other nuclear explosive device". This definition follows the usage in the NPT insofar as it rests upon a nuclear explosive capability irrespective of its intended military or peaceful function. It does not follow the usage in the NPT in that it does not restrict the acquisition of NWS status to the period before 1 January 1967. While this time limitation is functional for the purposes of the NPT, and would, of course, exclude India from the class of NWS under the treaty, it is unnecessarily restrictive in dealing with the broader issues involved in the long-term minimization of nuclear proliferation. Thus, utilizing this definition there are now six NWS: the USA, the USSR, the UK, France, the People's Republic of China and India. Clearly this class of states includes states with very disparate capabilities: the two major nuclear powers;[1] the three minor NWS; and India, which has tested only one nuclear explosive device for purportedly peaceful purposes and thus claims to have not yet exercised its nuclear weapon option.

The class of non-nuclear weapon states (NNWS) of interest here, often called near-nuclear or threshold nuclear weapon states, is defined as those states which are now technically and industrially able to develop the capability to manufacture and explode a nuclear weapon or other nuclear explosive device or will become so by the year 2000. This potential capability is already quite widespread and will become much more so by the year 2000. Fissile materials are becoming widely available as a by-product of industrial development and, in particular, of the increasing reliance throughout the world on nuclear energy to generate electric power. Assuming that alternative energy sources will not be widely exploited, the shift to nuclear energy for electric power generation will result in an annual world plutonium output from nuclear power reactors which may reach millions of kilograms by the year 2000. Nuclear fuel cycle facilities for uranium enrichment, fuel fabrication and chemical reprocessing are still quite concentrated in states with large nuclear power industries. However, such fuel cycle facilities are now being constructed more widely and technological innovations providing much lower cost processes will spur this trend [1, 2a, 3, 4].

The dual nature of nuclear fission is shown by the fact that between five and ten kilograms of plutonium are required to build an explosive device capable of doing significant damage to a medium-sized city. The proliferation of sensitive nuclear fuel cycle facilities or the capacity to build them thus also constitutes the proliferation of capabilities to develop nuclear weapons; in the nice phrase of Ernst Bergmann, the former Director of the Atomic Energy Commission in Israel, "There are no two atomic energies"[5]. Thus, as a NNWS acquires a nuclear power industry, it moves along what George Quester has called the "innocent progress toward the bomb" curve, in which the time lag between innocent progress in the development of a nuclear power industry and a crash programme to produce nuclear weapons sinks asymptotically to a very short time required for

[1] In this book, reference will often be made to the USA and the USSR as the 'two major NWS'. In other strategic and political writings on this subject, the usage 'nuclear superpowers' is often adopted to refer to the same two states.

residual efforts to fabricate at least a few crude nuclear weapons[6]. In effect, any state which has a small nuclear power industry, the sophisticated industrial infrastructure and skilled manpower which are the preconditions of a nuclear power industry, or the political will to divert the resources necessary to create such an industry, regardless of high opportunity costs, has the potential to become a NWS by the year 2000. This definition includes at least 30-40 states: great powers, middle powers and some small powers with high political incentives.[2] It excludes other small powers and all microstates, unless an international 'grey market' were to emerge in nuclear explosive devices which could make nuclear weapons easily available to such states[18].

Several implications of these definitions should be noted. First, it is clear that the class of NNWS of interest for this analysis ranges from states with major, advanced nuclear industries, such as FR Germany or Japan, to states with very primitive nuclear capabilities, such as Egypt, Pakistan or Yugoslavia. Furthermore, differences in potential delivery vehicle capabilities further extend the range of variation incorporated in this class. NNWS might possess delivery capabilities with local, regional or trans-regional ranges and varying in penetration capability, survivability and targeting flexibility. Thus the class of relevant NNWS includes states with a broad gamut of potential nuclear weapon capabilities.

The distinction between NWS and NNWS, given the large number of states which will have the potential to become NWS in the period under discussion, is somewhat arbitrary. NNWS remain non-nuclear under this definition if they make the political decision not to demonstrate their nuclear capabilities through publicly exploding a nuclear device. Simulation techniques are now so reliable that a state which placed high political value on the benefits of secrecy and wished to avoid the political costs of a public explosion might well forgo the demonstration of a nuclear explosive capability. Israel — which is widely reputed to have constructed or to be able to construct within a very few hours several nuclear weapons — and South Africa are cases in point[19]. The definition used here arbitrarily treats such states — which might be termed 'ambiguous' or 'non-demonstrated' NWS — as NNWS.

Finally, this definition of NNWS refers both to authoritative decision-making units which are now in power and to political groupings which may gain power before the year 2000 in the relevant NNWS. Such potential governments must include, in turn, all important political factions and relevant sub-state bureaucracies within the existing political structure as well as important competitive counter-élites, such as significant

[2]Collectively, various lists[2a, 2b, 2c, 7a, 8, 9a-17] suggest the possibility of 30-40 countries having nuclear capabilities by the year 2000. Countries which appeared on two or more of these lists were Argentina, Australia, Austria, Belgium, Brazil, Bulgaria, Canada, Chile, Czechoslovakia, the Democratic Republic of Germany, the Federal Republic of Germany, Finland, Greece, Hungary, Indonesia, Iran, Israel, Italy, Japan, Mexico, the Netherlands, Pakistan, the Philippines, Poland, Romania, South Africa, South Korea, Spain, Sweden, Switzerland, Taiwan, Thailand, Turkey, the United Arab Republic and Yugoslavia.

national liberation or separatist movements. It excludes non-state actors such as small revolutionary or terrorist organizations with no chance of gaining power in the relevant states. It also excludes profit-oriented criminal groups.

Clearly, the range of actors incorporated in the class of NNWS is an extremely broad one. In considering this range of actors, it is important to note that, by definition, the technical and industrial capabilities required to develop at least some nuclear weapons by the year 2000 are available to the NNWS. Furthermore, we can envisage no politically realistic export policies or safeguard systems which could prevent the government or potential government of a relevant NNWS from acquiring these capabilities and diverting nuclear materials for weapon purposes if it chose to do so.

Certainly, varying international export policies and international safeguard systems can impose different constraints upon the capabilities of various NNWS. Indeed, these are important policy instruments and are the focus of much current attention[1, 4, 20-25]. This is due in large measure to the growing interest of Third World NNWS in nuclear energy for electric power generation since the October 1973 War in the Middle East and the ensuing energy crisis. Many of these less developed states are only now beginning to acquire nuclear energy capabilities. Thus the rate and conditions of their acquisition of such capabilities can be controlled and regulated if the industrial states which are the major suppliers of fissile materials and nuclear facilities can agree upon appropriate policy mechanisms.

However, while the spread of nuclear energy capabilities can be regulated and perhaps retarded, it cannot be stopped, given the increasing reliance throughout the world on nuclear energy to generate electric power. Thus, over time, nuclear energy capabilities will gradually spread and policy constraints on such capabilities will be of dwindling efficacy. Thus, this book makes the simplifying assumption that there are no long-term means of *denying* NNWS the capabilities to acquire nuclear weapons if they choose to do so.

Therefore, if, in the long term, access to the necessary physical capabilities cannot be denied, and if the proliferation of nuclear weapons is to be minimized, the only recourse is to influence the intentions of NNWS with respect to their acquisition of independent nuclear weapon capabilities[26a]. This book will therefore not deal with policies to constrain the spread of nuclear energy capabilities. Rather, it will concentrate upon the long-term arms limitation and security policies which might satisfy the various objectives of NNWS, thereby minimizing whatever intentions they may have to acquire independent nuclear weapon capabilities and collectively minimizing future nuclear proliferation.

II. Premises

The destabilizing effects of nuclear proliferation

This book assumes that minimizing the proliferation of nuclear weapons to the range of NNWS just described is, over the long term, a necessary condition for a stable international security system. Indeed, it is assumed that the exercise of a nuclear weapon option by even one of the NNWS would be to at least a small degree destabilizing for the international security system. However, the optimal aim of preventing all future proliferation of nuclear weapons by the year 2000 must be modified by realistic assumptions about the amount of change possible in the international security system during this period. Barring major globe-threatening catastrophes inducing major shifts in international politics away from a nation-state dominated system, and given the resultant existence of independent nation-states during a period in which the capability to develop nuclear weapons will become increasingly widespread, it is realistic to expect that some proliferation may occur. To quote Hedley Bull on this point:

> In considering how far non-proliferation is feasible it is necessary to distinguish between stopping the spread of nuclear weapons and controlling it. It has never seemed likely at any point in the nuclear era, and it does not seem likely now, that all further proliferation will be stopped. It is simply not credible that one of the most vital strategic and political instrumentalities of the time, which is technically within reach of many states, will remain permanently the monopoly of the few that first developed it[27].

The purpose thus becomes one of designing a comprehensive arms limitation and security régime which can minimize nuclear proliferation, rather than prevent it altogether[28].

Minimizing the future proliferation of nuclear weapons to the governments and potential governments of NNWS is a crucial international policy objective because such proliferation may have a variety of destabilizing and high risk effects. Some analysts dispute this view, holding that in certain strategic situations, the acquisition of nuclear weapons by one or more adversary states might actually contribute to a relationship of stable, mutual deterrence, much as the development of invulnerable, second-strike nuclear weapon capabilities have arguably stabilized relations between the USA and the USSR[29-31]. Thus, such analysts conclude that in these instances, nuclear proliferation is either positively beneficial for, or at least not detrimental to, the stability of the international security system. Despite the superficial plausibility of this analogical reasoning, this argument is rejected here, since the possible risks inherent in the future proliferation of nuclear weapons are several. Ten of these risks, which are widely referred to in the strategic literature on nuclear proliferation, will be noted.

First, as a threshold condition, a domino effect may operate in which the emergence of a seventh or eighth NWS may encourage several more

NNWS to acquire a nuclear weapon capability. As will be noted in chapter 2, this is particularly true given the linkages between specific regional adversaries or competitors.

Second, with further proliferation of nuclear weapons, particularly the proliferation of primitive capabilities without reliable command and control procedures, the probabilities of accidental or unauthorized use of these weapons may increase.

Third, with further proliferation, future local wars between two middle or small powers may become nuclear wars. This is particularly true in the immediate post-proliferation transition period. In these early stages, the nuclear capabilities of one new NWS would be in their most primitive state and thus would be most vulnerable to nuclear or conventional attack either by the existing NWS or by regional NNWS adversaries. Alternatively, if each of two states which are regional adversaries acquires primitive and vulnerable first-strike nuclear capabilities and each becomes fearful of surprise attack, they may each be tempted to pre-empt.

Fourth, further proliferation may increase over time the probability of major nuclear war, in that local wars, in which nuclear weapons are first used, may lead to nuclear intervention and escalation by major NWS.

Fifth, further proliferation may increase the probabilities of catalytic war.

Sixth, any future use of nuclear weapons may, by breaching the post-war taboo against their use, render this firebreak a less effective deterrent against any subsequent use.

Seventh, further nuclear proliferation may provoke a continued arms race between the major NWS, since each might augment its own deterrent capability in order to be able to retaliate against attacks by all other NWS. More generally, it would make more difficult all future negotiations on arms limitation.

Eighth, further proliferation could be transferred to states, or counter-élites which are potential governments in states, which Yezhekial Dror has pungently characterized as "crazy"[8, 32]. Such crazy states, or crazy counter-élites within states, may be tempted to use nuclear weapons in non-normal, non-routine situations and in particularly nasty fashions. Such crazy states must be of serious concern since they are increasingly likely in a world in which many states or counter-élites within them have failed to achieve minimal aspiration levels, have intense feelings of deprivation, repression and injustice, and are disillusioned with contemporary values[33].

Ninth, while there are great difficulties in assessing the significance and likelihood of any of the above-mentioned risks, it is responsible to adopt 'worst plausible case assumptions' concerning the intentions of present and future decision-makers in NNWS. Analysts should be particularly careful in assessing the risks of proliferation in the long-term future. To say that we do not feel particularly threatened if State X acquires a nuclear weapon capability now is to make an optimistic prognosis about all potential governments and competitive counter-élites in State X which might inherit an organizationally established and, hence, probably irreversible nuclear weapon capability. Given the strands of militance and xenophobia and the important national-liberation or separatist movements which exist in many contemporary political cultures, such a prognosis may well be unwarranted.

Tenth and last, one may argue about the significance and likelihood of any one of the foregoing risks of proliferation, and even assert that in particular cases nuclear proliferation may not be destabilizing. The composite list of risks, however, is necessarily worrisome because the proliferation of nuclear weapons will in some such ways complicate international politics. If, in particular, several new NWS were to emerge in quick succession, the international security system might simply be swamped by the new level of complexity in international politics. Leonard Beaton cogently summarized these complicating effects of nuclear proliferation:

> This new complexity in the relations of powers is the most fundamental danger in a nuclear spread and, as so often appears to be the case with these weapons, the advantage will be with the incautious or apparently incautious.
>
> We cannot expect to have a clear idea of what a world with ten or twenty nuclear powers would be like. It might turn out, for example, that this large array of nuclear powers would feel an urge to create some sort of collective mechanism to control nuclear weapons. Other optimistic projections could be made. Nevertheless, the common instinct of revulsion at the prospect of a wide spread of nuclear weapons is undoubtedly the best guide. Many owners means many with the opportunity to use them or threaten their use; it means that many will be threatened by them; it means there will be difficulty in knowing who has been responsible for particular actions if these ever happen; it probably means measures of control will be more difficult to devise and negotiate; and it introduces uncertainties into international bargaining which might well be more than the human process of diplomacy and the complexities of alliances could sustain. A breach in the virtual nuclear monopoly of the Soviets, Americans and certain East Europeans might be equitable and even realistic; but the instinct which says that it makes the use of these weapons more likely cannot easily be denied . . . .
>
> If means are not found to contain proliferation, the whole structure of world security is going to become very difficult to sustain[34a, 35a, 36a-36c].

Discrimination in the international arms limitation and security régime and in the NPT

The NPT and the broader arms limitation and security régime in which it is embedded cannot minimize the proliferation of nuclear weapons — a necessary condition for a stable international security system — in large measure because it currently discriminates in favour of the particular interests of the NWS and against the particular interests of many of the NNWS. Indeed, the concept of 'discrimination' pervades discussion of the non-proliferation of nuclear weapons. Discrimination, when used as what lawyers term a 'word of art' in United Nations disarmament negotiating bodies, specifically refers to a disproportionately inequitable allocation of obligations and privileges among different states subject or potentially subject to an international negotiation or agreement concerning arms limitation and disarmament. The NPT has in many ways been perceived as the most visibly discriminatory of the post-war arms limitation agreements. However, concern with discrimination arises in connection with many other arms limitation and disarmament issues: in negotiations concerning nuclear testing and the control of chemical weapons; in aborted proposals for conventional arms control; in the general question of international access to national means of verification; and in the organization of ongoing arms limitation and disarmament negotiations. Thus, several general comments concerning the inevitability, the legitimacy and the significance of discrimination as it applies to arms limitation and disarmament should be made before specifically addressing the proliferation of nuclear weapons and the discriminatory nature of the NPT.

First, some differential allocation of capabilities is an inevitable fact of international life. Clearly the legal doctrine of the sovereign equality of states, upon which much of the formal organization of international society is based, runs afoul of the political facts of life. Some states are inevitably larger, richer and more powerful militarily than others.

Such disparities in the general strength of parties to a negotiation may cause them to adopt different negotiating strategies. Barbara Haskel, in an interesting study of the Scandinavian economic market negotiations, has concluded that

> there is a typical strategy of the strong which is likely to emphasize possible joint gains, reflecting confidence that the division of that gain will be either sufficient, or fair, or in their favor. Likewise there will be [a] tendency for the generally weaker to adhere to a distributive strategy, emphasizing their share of the joint gain, reflecting a lack of confidence as to whether anything not specifically allocated to them in advance will ever come their way later[37].

International politics have recently been characterized by this pattern of negotiation. Various cleavages have arisen between strong and weak, or 'Have' and 'Have-Not' states in such issue areas as environmental protection; access to energy-producing resources; the allocation of food, commodities and natural resources; economic development; and the evolving law of the sea.

A comparable cleavage, cross-cutting these basically North-South cleavages, has also developed in arms limitation and disarmament between what might be called the 'Military Have' and the 'Military Have-Not' states. This cleavage divides the two great powers and some close military allies from a wide range of other states, which include both industrialized and underdeveloped states; such 'western' states as Canada, Japan, the Netherlands and often France; such 'socialist' states as Romania and often China; and such non-aligned states as Argentina, Brazil, Egypt, India, Mexico, Sweden and Yugoslavia. Since military capabilities constitute the quintessential determinant of status in the international security system and since the USA and the USSR possess such pre-eminent military—and particularly nuclear—capabilities, domination by the 'Military Have' states in the arms limitation and disarmament issue area has been perhaps more long-lived and resistant to change than domination by 'Have' states in other issue areas. However, multilateral arms limitation and disarmament negotiations held since 1963 have increasingly manifested bargaining patterns similar to other issue areas in which the interests of the 'Have-Not' states are being pitted against those of the 'Haves'.

The 'Military Have' states, and particularly the United States and the Soviet Union, have, since 1963, rather cautiously pursued their own limited adversary relationship of détente. They have acquired increasingly sophisticated mutual deterrent capabilities and appear committed to maintaining their unique status as the major nuclear weapon states constrained only by certain mutually acceptable partial measures of arms limitation or non-armament. They have avoided serious negotiation on nuclear disarmament and have favoured bilateral negotiations or joint domination of multilateral negotiations.

On the other hand, the 'Military Have-Not' states, despite differing national security objectives, have since 1963 attached high priority to the objective of substantial nuclear disarmament within a short period and sought greater participation in multilateral negotiating bodies responsive to their interests as well as to those of the USA and the USSR. To quote Alva Myrdal, the former Ambassador of Sweden to the Eighteen Nation Disarmament Committee/Commission of the Committee on Disarmament (ENDC/CCD), on this point:

> During most of the post-war period, the field of vision of each of the superpowers has been restricted to the narrow one of its national interest vis-à-vis the other. But in disarmament nothing can be done if one does not consider the broader, long-term interests of the world, where one's own security is part of the more general picture.
>
> Evidently the superpowers have had a joint interest in preserving their respective positions and their mutual balance, and in not letting it be disturbed by any disarmament measure for which other nations might be working. The consequence has been that the disarmament negotiations have been used by the superpowers for balancing each other and not for planning disarmament [38, 39].

The fact that the preferences of the major nuclear powers have consistently dominated the preferences of the NNWS is perceived by the latter as a *prima facie* case that arms limitation and disarmament negotiations and agreements have to date discriminated in favour of the particular interests of the stronger NWS and against the particular interests of the weaker NNWS.

Second, the resultant perceived discrimination, while to some degree inevitable, is, nonetheless, problematic, for it raises important normative questions about the legitimacy of contemporary arms limitation and disarmament negotiations and agreements. The NWS, particularly the

United States and the Soviet Union, should not assume that the inevitably differential allocation of military capabilities in the international security system entitles them to disregard the attitudes of weaker states concerning the limitation of these capabilities. On the contrary, the attitudes of weaker states, particularly NNWS, toward nuclear weapons may well constitute a more legitimate set of standards by which to identify the common interests of the international security system than the major nuclear powers have been willing to grant since 1945. To quote Richard Falk, writing in 1963 on this point:

> It is usual . . . to discount Afro-Asian hostility to nuclear weapons as irresponsible criticism from uninformed sources. This strikes me as a supercilious and complacent response from the West, even though it may be sufficiently accurate as a certain kind of explanation. Detachment, even alienation from power, often produces irresponsible positions . . . but it may also produce a juster sense of proportion, a more discriminating and objective priority scale. We accept this as conventional wisdom in the context of dispute-settlement and judicial administration; it hardly occurs to anyone that the partisan would make a better judge than the impartial observer. In similar fashion, I would urge that we should not be quite so quick to repudiate the priority scale recommended by the non-nuclear states or quite so confident about the priorities of the nuclear states. Neutral attitudes about the illegitimacy of nuclear weapons deserve attention.
>
> . . . It is said that there are no non-partisan states in international politics, that the aversion to nuclear weapons represents an attempt by non-nuclear powers to make up for their military insignificance, and that only states with a direct interest in the outcome of political controversies are in a position to make a rational calculation of the utility and disutility of a controversial policy. Such an attitude is partly accurate, partly simplistic. There are a multitude of factors that combine to produce a political viewpoint. Access to power, although an important one, is not the necessarily dominant factor and it is certainly not the only one . . . States with insignificant power may have a better understanding of the limits of national power in the nuclear age than stalemated nuclear states who also possess insignificant usable power, but are slower to shed illusions about the adequacy of national power because of their enormous destructive potential[40a, 41].

In other words, it is not self-evident that NWS have a legitimate prior claim to make authoritative and presumably self-interested decisions concerning the limitation of nuclear weapons. In designing policy by which to achieve an arms limitation and disarmament objective such as minimizing the proliferation of nuclear weapons, charges by the NNWS that current policies are discriminatory should at least be considered legitimate, and be seriously addressed.

Third, the discriminatory character of certain arms limitation and disarmament negotiations and agreements is significant since, for at least two reasons, it renders them inherently unstable over time. Agreements which permit some stronger states disproportionately inequitable privileges which are denied to weaker states cannot easily be used to impose moral pressure on states refusing to comply. Thus, these agreements may be unstable because adherence to them is difficult to marshal. Both the Partial

Test Ban Treaty, in the case of France and China, and the NPT, in the case of many relevant NNWS which have refused to sign and/or ratify, are examples of such agreements. Furthermore, discriminatory agreements can be sustained only so long as there are important security relationships between those states which the agreement favours and those it inhibits[42]. Unless compensatory counterdiscriminatory obligations are undertaken by those states which the agreement favours, the agreement is likely to be ineffectual or, if effective in the short term, to deteriorate over time.

Turning specifically to the proliferation of nuclear weapons, the differentiation of states into the two classes of NWS and NNWS, each subject to different prohibitions and regulations, has not been viewed as *ipso facto* discriminatory. Obviously, such a distinction is inherent in any attempt to control the proliferation of nuclear weapons by freezing acquisition of such weapons to a limited class of states. What is considered discriminatory in the NPT, in addition to issues concerning the export and safeguarding of sensitive nuclear materials and facilities which are not addressed here, is the distribution of obligations, risks, expectations and privileges in the broader arms limitation and security régime as they differentially affect first, the military security and second, the political prestige of these two classes of states. As discussed below, the NPT asks NNWS to forgo an extraordinary sovereign right: the right to pursue their military security and political prestige objectives through the manufacture —*even from their own resources*—of the most sophisticated military weapons extant. On the other hand, the NPT in no way constrains the right of NWS to pursue their military security and political prestige objectives through the development, testing, production, deployment or use of such weapons[38, 39]. It is widely felt by the NNWS that the NWS have not compensated for these extraordinary privileges by undertaking and executing commensurate counterdiscriminatory obligations relating to arms limitation, disarmament and the maintenance of an international security system which will fulfil the various objectives of NNWS.

Nuclear proliferation in a comprehensive and global perspective

It is assumed here that one cannot usefully analyse the NPT as a separate arms limitation measure which, standing alone, is designed to constrain the proliferation of nuclear weapons. Rather, the NPT must be viewed as one component of the much more comprehensive international security régime which encompasses the arms acquisition, arms limitation and security policies of both the nuclear weapon states and the non-nuclear weapon states[43]. If the NPT and the comprehensive international security régime in which it is currently embedded cannot effectively minimize the future proliferation of nuclear weapons, and if one is interested in minimizing the future proliferation of nuclear weapons, one must identify additional, mutually compatible arms limitation and security policies, affecting both

NWS and NNWS, which will buttress and amplify the NPT. Such a comprehensive anti-proliferation régime must satisfy the various objectives of NNWS which might, alternatively, be satisfied by their acquisition of independent nuclear weapon capabilities[44]. Otherwise, the NPT will be degraded and, much more importantly, the proliferation of nuclear weapons will continue.

In attempting to identify the long-term arms limitation and security policies which might satisfy various objectives of NNWS and thus collectively minimize future nuclear proliferation, it is important to seek to maximize the collective, common interests of the international security system rather than the security interests of particular nation-states. This distinction between the international system on the one hand and the nation-state actor on the other represents a classic dichotomy in the level of analysis at which one studies international politics. The analysis in this book is grounded in the collective perspective of the international security system shared by such scholars as Leonard Beaton, Hedley Bull, Richard Falk, Saul Mendlovitz, J. David Singer and Martin Wight[41, 42, 45-48].

Alastair Buchan aptly elucidated this collective perspective as it applied to the issue of nuclear proliferation in 1966 when he invoked Jean Jacques Rousseau's fable of the hare and the stag:

Five men living in a "state of nature" as nations still do, converge at a time when they are all hungry: they agree that their hunger will be assuaged by one fifth of a stag and they lay a trap to encircle one: but a hare, sufficient to appease the hunger of any one of them, dashes by and one man grabs it: his hunger is satisfied but the stag escapes. Rousseau used this fable to illustrate the conflict between the particular and the general interest which, in his view, was the basis of international conflict. Although Rousseau wrote nearly two centuries before nuclear weapons became a reality, his fable has an even clearer implication today than in his own or succeeding generations. Each of the five existing nuclear powers has, at one time or another, acted as the man who impaired the hope of collective action to prevent the spread of nuclear weapons by grabbing the hare.... Each government has been able to produce reasons to convince the majority of its own people why the improvement of its own national security should have priority. It has been able to appease consciences by the assertion that the certainty of the hare offers a better immediate prospect of survival and stability than foregoing it in the uncertain prospect of bigger game.

...For nearly a decade it has been apparent that the arguments that have led to the final decision to remain or become a nuclear power in London, Paris or Peking may apply with equal force to other countries with similar problems or ambitions to those of the original five; and that unless conscious steps are taken, either to remove the incentives or to control the means, the number of hare snatchers might be doubled in a decade, and doubled or trebled again by the end of the century, while the stag has disappeared forever[35b].

This collective perspective has been somewhat ignored in much recent work on nuclear proliferation which tends to emphasize comparatively short-term, concrete, political calculations by which particular nation-states and sub-state bureaucracies within particular nation-states determine whether or not to exercise the nuclear weapon option. This literature tends to discount abstract, systemic analyses of nuclear proliferation to '*n*th countries'[5, 49, 50]. The collective perspective is, however, an important alternative one from which to view the issue of nuclear proliferation since, as the fable of the hare and the stag suggests, it corrects for certain shortcomings in the disaggregated nation-state perspective.

To put the argument somewhat differently, economic theory suggests that decisions by sub-units within a total system tend to be sub-optimal for the total system under conditions where important externalities exist. It is clear in considering the proliferation of nuclear weapons that there are important externalities for the international security system when a particular nation-state calculates whether or not to exercise its nuclear weapon option. Positive economies may arise for the system if a state chooses not to exercise its option and, conversely, diseconomies may arise if it does choose to exercise it.

This is not to state that the calculations of national interest by particular nation-states which do not take such externalities into account are in any sense illegitimate. It does assert, however, that such calculations may be sub-optimal for the international security system as a whole. Nation-states as entities have 'national interests' — however determined and however diluted or skewed by bureaucratic politics — and national decision-makers who are responsible to national constituencies are in some sense obliged to pursue them. These national interests, however, remain particular interests, and they are not likely to be the equivalent of the collective interest of the international security system. In Leonard Beaton's words: "Each nation has its own security policy and its own interests. But the world as a whole has interests: and as in any larger political entity constructed from lesser but self-conscious units, the general interest supports some elements of each particular interest and opposes or confines others"[42a].

Therefore, given the discrepancies which may arise between the collective and the particular interests involved in a nation-state's decision whether or not to exercise its nuclear weapon option, a collective perspective provides standards by which to design the mutually compatible arms limitation and security policies most likely to satisfy the various objectives of NNWS and thereby minimize the future proliferation of nuclear weapons. It also provides standards by which to determine whether the decisions of particular nation-states at a given point in time conform with these policies. The identification of such policies is an important function of independent scholars and research organizations concerned with issues of arms limitation and international security since representatives of nation-states, as custodians of particular interests, cannot qualify as what Hedley Bull has rather grandly called "legislators of the common good"[43, 51a].

2. Arms limitation and security policies for minimizing nuclear proliferation: the strategic debate

This chapter will review the debate in the strategic literature on nuclear proliferation concerning the arms limitation and security policies which might best buttress and amplify the NPT and minimize the proliferation of nuclear weapons by the year 2000. The discussion is divided into three parts. First, the military security and political prestige objectives of NNWS which must be satisfied by a broad anti-proliferation régime if nuclear proliferation is to be minimized in this period are reviewed. Second, two hypothetical, composite arms limitation and security strategies which the NWS might adopt in order to meet various objectives of the NNWS are described. These two strategies — the High Posture Doctrine and the Low Posture Doctrine — have been considered, since discussion of nuclear proliferation began in the late 1950s, as the principal alternative means by which the NWS might influence the intentions of NNWS with respect to their acquisition of independent nuclear weapon capabilities. Third, the proposed arms limitation and security policies which best satisfy the various objectives of NNWS and are thus appropriate for amplifying the NPT and minimizing the future proliferation of nuclear weapons are identified. It is concluded that a modified Low Posture Doctrine — one which includes residual positive security guarantees for use against actual or threatened nuclear attack — constitutes the comprehensive anti-proliferation régime which is most likely to minimize future nuclear proliferation.

I. Policy objectives of NNWS

As defined in chapter 1, some 30-40 NNWS have or by the year 2000 will have a nuclear power industry or industrial base and thus the technological capability to exercise a nuclear weapon option. Such states may choose to exercise this nuclear weapon option in order to enhance their military security and/or their political prestige relative to other important referent states. Thus, in order to minimize future nuclear proliferation, a comprehensive anti-proliferation régime must include mutually compatible arms limitation and security policies which will satisfy the objectives of various classes of NNWS at least as effectively as would their acquisition of independent nuclear weapon capabilities.

Four specific military security and four specific political prestige objectives of NNWS are identified in the two sub-sections below. A given NNWS might have several such objectives which varying arms limitation and security policies would have to satisfy if the state is to continue to forgo the acquisition of an independent nuclear weapon capability. Seven geopolitical regions can also be identified in which nuclear weapons might proliferate by the year 2000: South Asia, the Middle East, East Asia, Australasia, Southern Africa, Latin America and Europe[9b, 52-54]. For each of the two types of objective under consideration, an attempt is made to identify all plausible examples of NNWS which might choose to acquire nuclear weapons in order to achieve the given objective. A summary is outlined in table 2.1 (p.22). Although some of these examples may now appear to be extremely low-probability contingencies, they may become more likely or may even have materialized by the year 2000.

It should also be noted that various coalitions of domestic political and bureaucratic-organizational interests might account for any given NNWS's decision to acquire nuclear weapons. Some component of the military and technical communities must favour exercising the nuclear weapon option for it to go forward, but the bureaucracy and the wider political system are not likely to be uniformly favourable to the acquisition of nuclear weapons. Thus, possibilities exist for external manipulation of the internal decision processes concerning nuclear weapon acquisition. These internal decision processes must be analysed on a case-by-case basis, so they are not enumerated here. However, in discussion of the policy objectives of NNWS as if they were unitary actors, it is important to recall that in fact various arms limitation and security policies are likely to be differentially influential with important bureaucratic and political interests in any given NNWS[55].

Military security objectives

First, NNWS may seek to enhance their military security through the acquisition of nuclear weapons. With the exception of the Sino-Soviet fighting on Chen Pao/Damansky Island in March 1969, no NWS has ever had its homeland placed under any form of attack by any nation-state adversary after having acquired nuclear weapons. Historically, nuclear weapons have thus appeared to be an effective military deterrent. If a NNWS has a pressing concern for its military security, acquiring nuclear weapons clearly becomes a salient policy option.

There are four specific military security objectives for which the acquisition of nuclear weapons could seem an appropriate option for NNWS: *first*, deterrence of, defence against and/or retaliation for a nuclear or conventional attack or nuclear blackmail by the USA or the USSR; *second*, deterrence of, defence against and/or retaliation for a nuclear or conventional attack or nuclear blackmail by a minor NWS; *third*, deterrence of, defence against and/or retaliation for conventional attack by a neighbouring or regional adversary NNWS or group of NNWS adversaries or domination of such NNWS adversaries; and *fourth*, anticipatory reaction to the prospective acquisition of nuclear weapons by a local or regional NNWS adversary in order to deter or dominate such an adversary. Taking each of these four objectives in turn, all the plausible scenarios in which NNWS might choose the nuclear option to attain the respective objective are enumerated below.

Objective 1. The USSR, the UK, France and China acquired substantial nuclear weapon inventories, in order, among other reasons, to deter nuclear attack or blackmail by a major nuclear weapon state. Furthermore, the UK, France and NATO, according to the doctrine of flexible response, rely upon a policy of possible first use of nuclear weapons in order to deter or defend against a possibly superior conventional attack by the USSR and its Warsaw Treaty Organization (WTO) allies[26a, 26b, 56a].

The force structures which NNWS might acquire in order to achieve this objective will vary with the state's size and wealth, as measured in terms of some combination of population and gross national product (GNP). Regardless of the particular force structure, all such states can be viewed as attempting some form of 'proportional deterrence' by threatening the major nuclear power adversary with an unacceptably high risk of damage to its population or industrial base compared with the strategic value of the state in question.

A great power might attempt to develop a second-strike force sufficiently large and invulnerable to deter strategic attack or blackmail by its major nuclear power adversary. It might, in addition, attempt to develop tactical nuclear weapons in order to defend against conventional attack by its adversary. NNWS which might acquire such strategic and/or tactical capabilities in order to achieve this objective include: FR Germany, Japan and a future Western European Community, should one come into existence, against the USSR and its WTO allies.

A middle or small power might attempt to develop a tactical nuclear weapon force sufficient to deny its adversary military victory at an acceptable level of costs and risks. NNWS which might acquire tactical nuclear weapons in order thus to deter or defend against a nuclear or conventional attack by a major nuclear power adversary include: Austria, Belgium, Canada, Finland, Greece, Iran, Italy, the Netherlands, Romania, Sweden, Switzerland, Turkey and Yugoslavia against the USSR and its WTO allies[57, 58]; and Bulgaria, Czechoslovakia, the German Democratic Republic and Poland against the USA and its NATO allies.

Finally, a middle or small power might acquire a small, vulnerable, preemptive first-strike force in order to make suicidal deterrent threats and thereby create uncertainties for a hostile major NWS or, *in extremis*, actually to retaliate against such an adversary's threat or mounting of a conventional attack. Such a capability would imply that the country, faced with immediate destruction, would prefer "suicide at a high cost to the adversary to loss of the conflict"[8]. NNWS which might acquire such nuclear weapons in order to deter or retaliate against such a conventional attack include Israel, Romania and Yugoslavia against the USSR, and Cuba and conceivably Libya and Saudi Arabia against the USA.

Many of these NNWS, including great, middle and small states, now depend on fairly stable, explicit or tacit positive security guarantees from one of the major NWS in order to meet the objective of deterring or defending against nuclear or conventional attack by the other. Such NNWS might consider exercising their nuclear weapon option should such positive security guarantees lose their credibility and should no alternative security policies be forthcoming.

Objective 2. India may be developing such forces, among other reasons, in order to achieve the second type of military security objective against China: namely, deterrence of, defence against, and/or retaliation for a nuclear or conventional attack or nuclear blackmail by a minor NWS. This objective might be sought by great, middle and small states utilizing a variety of nuclear force structures including an invulnerable second-strike force, a tactical nuclear weapon force, or a small vulnerable pre-emptive first-strike force. NNWS which might acquire nuclear weapons in order to achieve this objective include: Japan and Taiwan against China; Pakistan and Iran against India; and Australia and Indonesia against China and/or India. Of these states, Taiwan — increasingly an international pariah state — and potentially Pakistan face particularly difficult regional conflict situations and may feel that their very survival depends upon the acquisition of nuclear weapons.

Such contingencies clearly cluster into specific regional groupings. Thus the acquisition of nuclear weapons by China and India may provoke a local or regional 'domino effect' or 'chain reaction' in South Asia, East Asia and Australasia, whereby the NNWS adversaries and competitors of China and India proceed to acquire nuclear weapons. Some of these states have not received sufficiently credible positive security guarantees from one or more NWS and enjoy no alternative security policies which would provide effective substitutes for such positive security guarantees. Such NNWS thus have a strong incentive to acquire at least a tactical nuclear weapon capability or a small strategic nuclear force since, without any nuclear weapons, they are peculiarly vulnerable to nuclear or conventional attack or nuclear blackmail by their local or regional NWS adversary.

Objective 3. India may be developing a nuclear force in order, among other reasons, to achieve the third type of objective *vis-à-vis* Pakistan: namely, deterrence or, defence against, and/or retaliation for conventional attack by a neighbouring or regional adversary NNWS or domination of such NNWS adversaries. This objective might be achieved by nuclear force structures ranging from a tactical nuclear weapon force, a small vulnerable pre-emptive first-strike force, or an ostensibly peaceful nuclear explosive capability. It is particularly likely in the case where a NNWS faces hostile local or regional NNWS enjoying an existing or potential advantage in conventional military capabilities which, by dint of population size, natural resources and/or potential industrialization, is both overwhelming and irreversible. NNWS which might acquire nuclear weapons in order to achieve this objective include: Israel aganst a group of non-nuclear weapon Arab states; South Korea against a non-nuclear weapon North Korea; South Africa against a group of non-nuclear weapon Black African states, possibly supported by coercive states elsewhere; Iran against a non-nuclear weapon Iraq; and Australia against a non-nuclear weapon Indonesia.

Most of these states have, in the past, received fairly credible US guarantees of continuing conventional arms supplies or promised military intervention should a conventional attack occur. However, their faith in the future reliability of such guarantees is declining. Conversely, they may feel that the acquisition of an independent nuclear weapon capability might strengthen a major NWS guarantor's commitment to defend its client, lest an unprotected client feel impelled to use its nuclear weapons. Of these states, Israel, South Korea and potentially South Africa — increasingly viewed as international pariahs — face particularly difficult regional conflict situations and may feel that their very survival depends upon the acquisition of nuclear weapons. 'Crazy states' might pursue this objective as well. It should be noted that such decisions to exercise a nuclear weapon option may well generate the local or regional domino effect discussed in connection with objective 2, whereby the NNWS adversary is subsequently under enhanced pressure to acquire nuclear weapons. Objective 3 can thus also merge into objective 4.

Objective 4. There are no current NWS which acquired nuclear weapons in order to achieve the fourth objective: namely, as an anticipatory reaction to their prospective acquisition by a local or regional NNWS adversary. This objective might be achieved by a tactical nuclear weapon force; a small, vulnerable, pre-emptive first-strike force; or an ostensibly peaceful nuclear explosive capability. Any one of the several local or regional pairs or groupings of NNWS adversaries might acquire nuclear weapons as an anticipatory reaction to their prospective acquisition by an adversary in order to deter or dominate it. Such an action may well generate a counteraction by its adversary which would then, as discussed in connection with objectives 2 and 3, feel increasingly threatened by the new NWS and thus under enhanced pressure to acquire nuclear weapons. A local or regional nuclear arms race might then ensue, evincing local or regional domino effects. Should both states acquire small, vulnerable, pre-emptive first-strike forces, such an arms race would be quite destabilizing. Clusters of regional NNWS adversaries which might acquire nuclear weapons in order to achieve the fourth objective include: North Korea and South Korea against each other; Greece and Turkey against each other; Nigeria and Zaire against South Africa; Egypt, Iraq, Libya, Saudi Arabia and Syria against Israel or Iran, or against each other should any one of them appear to be acquiring nuclear weapons; Australia, Indonesia, the Philippines and Thailand against Japan or against each other should any one of them appear to be acquiring nuclear weapons; and Argentina, Brazil,

Chile, Cuba and Mexico against each other should any one of them appear to be acquiring nuclear weapons. 'Crazy states' might pursue this objective as well.

Political prestige objectives

In addition to, or in some instances in place of, such military security objectives, NNWS may choose to acquire nuclear weapons in order to enhance their political status and prestige. Enhancing political prestige is a serious objective for any state, since such increased status represents potential political power and influence which can be applied in many international situations.

Many NNWS observe that nuclear weapons have served important symbolic functions since the onset of the nuclear age. Like the classic role of gold in the international monetary system, nuclear weapons have come to constitute perhaps the quintessential attribute of prestige and determinant of status in the international political hierarchy[26c]. Nuclear weapons symbolize a state's modernity, scientific prowess and technological dynamism and thus clearly possess political as well as military utility. Whether out of perceived domestic or international political needs, should the government of a NNWS have a pressing concern for enhancing its political prestige, acquiring nuclear weapons is a salient policy option.

Again, there are four specific prestige objectives — objectives 5 to 8 — for which the acquisition of nuclear weapons might seem an appropriate option: *fifth*, enhanced political prestige with reference to the great powers in the post-war international political hierarchy; *sixth*, enhanced political prestige with reference to existing military alliances; *seventh*, second-order power status and/or enhanced political prestige with reference to a particular local or regional grouping of states or transregional status cohort in the international political hierarchy; and *eighth*, enhanced political prestige in order to change the existing distribution of status and power in the international political hierarchy. Here also, possible scenarios of NNWS seeking to reach each objective by nuclear means are presented below.

Objective 5. Some NNWS may seek to acquire nuclear weapons in order to achieve great power status — the highest status cohort in the existing international political hierarchy as measured by some combination of population, wealth and historical-cultural tradition in addition to military power. A nuclear weapon capability has, since the onset of the nuclear age, constituted an explicit component of great power status. Indeed, until India tested its nuclear explosive in May 1974, the five NWS were the only states with permanent membership in the UN Security Council, another explicit indicator of post-war great power status. The nuclear weapon capabilities of the USA, the USSR and the UK were derived from programmes originating in World War II and directed against the Axis Powers. France,

China and India — as well as the UK — more recently initiated nuclear weapon programmes in order, among other reasons, to lay claim to the manifest status of a great power and thus to join the 'top table' in the post-war international political hierarchy[26c]. Certain NNWS such as Brazil, FR Germany, Iran, Japan, South Africa and a future Western European Community, should one come into existence, might plausibly lay claim to such great power status in the existing international political hierarchy. Such NNWS might acquire nuclear weapons in order to rectify their perceived status inconsistency by conforming their military capabilities to their other components of great power status[59].

Objective 6. In addition to seeking great power status in the international political hierarchy, as in objective 5 above, France and China acquired nuclear weapons in order to achieve more political independence and higher political status in relation to their respective NWS allies. France and China each sought to transform their political alliances into less hegemonic relationships by, respectively, challenging the supremacy of the USA and the UK within NATO, and the USSR within the Sino-Soviet Treaty and Agreement of 1950[26c].

India, in pursuit of a related objective, may be developing a nuclear weapon capability in order, among other reasons, to assert its disinterest in joining any tacit military alliance with or receiving any positive security guarantees from existing NWS. India thus emphasizes its commitment to a non-aligned political status independent of existing military alliances.

Certain NNWS, such as FR Germany, Japan and a possible future Western European Community, might acquire nuclear weapons in order to gain enhanced political prestige with reference to the USA — the dominant NWS ally in these cases. Other NNWS, such as Brazil or Iran, might acquire nuclear weapons in order to assert their non-aligned political status independent of existing military alliances.

Objective 7. Apart from or in addition to seeking manifest status as a great power or enhanced status with reference to military alliances, some NNWS may acquire nuclear explosive capabilities or nuclear weapons in order to gain or reinforce what S.B. Cohen has termed 'second-order power status'. Such status would provide symbolic leadership with reference to a particular local or regional grouping of states or transregional status cohort in the international political hierarchy[60a]. India acquired a nuclear explosive capability in order, among other reasons, to achieve this composite objective. India has reason to expect that such a nuclear capability will buttress its status and leadership claims in the local subcontinent and enhance its status in the broader Indian Ocean region and the Third World — the more general status cohort which India has, at times, aspired to lead.

Now that India has exploded a nuclear device, other NNWS in the same region or status cohort may experience some relative loss of status. Should such NNWS perceive that their local, regional or transregional competitors, such as India, achieve second-order power status by exploding a nuclear device or acquiring nuclear weapons, they may be tempted to follow suit and an arms race for political prestige may ensue. Thus the pursuit of this composite objective may result in the type of pre-emptive regional nuclear arms races discussed in connection with military security objective 4, since acquiring nuclear explosive capabilities or nuclear weapons in anticipation of a competitor's doing so clearly may be justified on the grounds of political prestige as well as military security.

The NNWS which might at some time seek to achieve this composite objective include: Algeria, Argentina, Australia, Brazil, Chile, Cuba, Egypt, Greece, Indonesia, Iran, Iraq, Italy, Libya, Mexico, Nigeria, North Korea, the Philippines, Saudi Arabia, South Africa, South Korea, Spain, Syria, Taiwan, Thailand, Turkey, Venezuela and Zaire. Many of the foregoing states are in the less developed world and may have relatively few economic and political resources with which to gain political status. Such states, with few alternative sources of political prestige, may be under particular pressure to acquire or at least to threaten to acquire nuclear weapons so as to maximize their otherwise low levels of potential political influence and to gain political and economic benefits *vis-à-vis* their local or regional status cohort. 'Crazy states' might pursue this objective as well. Such competition for prestige and leadership, by definition, falls into specific local, regional or transregional clusters. Thus, if some nuclear proliferation occurs among such a group, the remaining NNWS may feel a relative loss of status and move to acquire nuclear weapons. In this way, the pursuit of second-order power status threatens to bring about the same domino effects discussed above in connection with military security objectives 2, 3 and 4.

Objective 8. Rather than raising their particular status in the existing international political hierarchy, some NNWS may seek instead to alter the hierarchy itself in the direction of greater equality among states. Should the proliferation of nuclear weapons become fashionable and its momentum build to such a point that a large number of middle and small powers acquire nuclear weapons, nuclear weapon capabilities will obviously cease to connote automatic great power or even second-order power status. Thus, if several middle and small powers perceive nuclear weapons as a 'great equalizer' and expect that by acquiring them they can collectively change the existing distribution of power and status in the international political hierarchy, they may have an added incentive to exercise their nuclear weapon options.

As noted in chapter 1, many states in the Third World perceive the current international political hierarchy as inherently discriminatory in favour of the great powers and other industrialized states and against their own interests. Such a discriminatory world order is rapidly becoming less viable. Dependence upon great powers is increasingly unacceptable to many states, and demands for equal status among states are voiced in many issue areas and institutions in contemporary international politics[60-64]. The issue area of nuclear proliferation is no exception. The widespread criticism

Table 2.1. The policy objectives of NNWS which might be achieved by acquiring a nuclear weapon capability

1	2	3	4	5	6	7	8
						Algeria	Many NNWS, including most of the states in the Third World listed here
			Argentina			Argentina	
	Australia	Australia	Australia			Australia	
Austria							
Belgium							
			Brazil	Brazil	Brazil	Brazil	
Bulgaria							
Canada							
			Chile			Chile	
Cuba			Cuba			Cuba	
Czechoslovakia							
			Egypt			Egypt	
German DR							
FR Germany				FR Germany	FR Germany		
Finland							
Greece			Greece			Greece	
Hungary							
	Indonesia		Indonesia			Indonesia	
Iran	Iran	Iran	Iran	Iran	Iran	Iran	
			Iraq			Iraq	
Israel		Israel	Israel				
Italy						Italy	
Japan	Japan		Japan	Japan	Japan		

Libya			Libya			Libya
			Mexico			Mexico
Netherlands						
			Nigeria			Nigeria
			North Korea			North Korea
	Pakistan					
			Philippines			Philippines
Poland						
Romania						
Saudi Arabia			Saudi Arabia			Saudi Arabia
		South Africa	South Africa	South Africa		South Africa
		South Korea	South Korea			South Korea
						Spain
Sweden						
Switzerland						
			Syria			Syria
	Taiwan					Taiwan
			Thailand			Thailand
Turkey			Turkey			Turkey
						Venezuela
Yugoslavia						
			Zaire			Zaire
Future Western European Community				Future Western European Community	Future Western European Community	

by many NNWS in the Third World of the discriminatory nature of the NPT, of its associated international safeguards and of various nuclear suppliers' policies, and their relatively positive reception of China's first nuclear explosion in 1964 and India's in 1974 attest to these anti-hegemonic attitudes towards the NWS and other traditional great powers. This reaction suggests that for many NNWS which advocate a redistribution of power and status in the international political hierarchy, the acquisition of nuclear weapons may appear to be a salient means of achieving this objective[27].

Moreover, once nuclear proliferation becomes widespread, NNWS with no particular drive to change the existing international political hierarchy, such as middle or small industrial powers, may feel impelled to acquire nuclear weapons merely to keep pace with the many other states which have already done so.

In summary, there are many possible policy objectives of NNWS which might be achieved by the acquisition of an independent nuclear weapon or nuclear explosive capability. As noted in table 2.1 (page 22), the most common categories are objectives 1 (deterrence of attack by a major NWS), 4 (anticipatory reaction to a regional adversary's acquiring nuclear weapons), 7 (enhancing political prestige by achieving second-order power status), and 8 (changing the existing international political hierarchy).

Most NNWS interested in achieving objective 1 are the industrialized European states which have for some years been the recipients of fairly stable and hence credible positive security guarantees from one major NWS guarantor against the other major NWS. These states, and the satisfaction of their military security objectives, were the original targets of the anti-proliferation régime evolved in the early and mid-1960s and culminating in the NPT.

In the past decade, however, numerous other NNWS have emerged which are interested in achieving objectives 4, 7 and 8. These states are, by contrast, primarily less developed states in the Third World which have not received, or do not want to receive, positive security guarantees from the NWS. In the past, most of these states have lacked the capabilities to exercise an independent nuclear weapon option. However, as noted in chapter 1, such technical constraints are diminishing as nuclear energy capabilities become more widespread. Thus, as we shall see in the subsequent sections, the requirements for designing a comprehensive arms limitation and security régime which will minimize the intentions of such states to acquire nuclear weapons by the year 2000 are becoming increasingly demanding.

II. Alternative strategies to satisfy policy objectives of NNWS

If the proliferation of nuclear weapons is to be minimized, a comprehensive anti-proliferation régime must include various mutually compatible arms limitation and security policies which can to some degree satisfy the military security and political prestige objectives of the NNWS enumerated above. Although such policies affect the entire international security system, they are largely subject to control and manipulation by the NWS, and in particular, by the USA and the USSR. Thus, it is assumed that some causal relationship exists between the arms limitation and security policies of the NWS and subsequent decisions by the NNWS whether or not to develop, acquire and deploy a particular nuclear force structure at a particular rate[65a, 65b, 66].

This relationship between the policies of the NWS and the nuclear decisions of the NNWS exists primarily because the NWS, particularly the USA and the USSR, currently occupy the dominant positions of power and status in the international security system. In general, the NWS now serve as the major 'custodians' of world order, and their behaviour in the international security system is an unavoidably important determinant of the policy choices of other nation-states. By their ongoing choices regarding such controllable policy variables as the size and quality of their own nuclear weapon inventories, the deployment of their nuclear weapon inventories, the permissible contingencies in which these capabilities might be used, the political utility of nuclear weapons in their conduct of foreign policy, and their broader security policies towards the NNWS, the NWS largely construct and manage the international order and climate within which NNWS must make their nuclear choice. There may be, as we shall note below, a few specific instances in which the decisions of particular NNWS to acquire nuclear weapons are insensitive to such policy choices of the NWS. The decisions of so-called international pariah states may be such instances. However, over time the NWS can and do consistently "influence perceptions about the utility and centrality of nuclear weaponry in the relations of states"[26c].

What then is the form of this assumed relationship between the policies of the NWS and the choices of the NNWS? In a seminal article published in 1967, Hedley Bull outlined two alternative, hypothetical, grossly linear relationships between the arms limitation and security policies of the NWS, as the independent variable, and nuclear proliferation by NNWS, as the dependent variable. Bull termed the two alternative models of NWS behaviour — each a strategy designed, among other objectives, to minimize nuclear proliferation — as the doctrines of 'High Posture' and 'Low Posture', respectively. The High Posture Doctrine and the Low Posture Doctrine each constitutes a comprehensive arms limitation and security régime. Each attempts systematically to regulate and manage the security environment of all NNWS rather than to influence each particular NNWS episodically, differentially and inevitably inconsistently on a case-by-case basis[53, 54, 65a].

These two hypothetical postures are, moreover, relatively polar models of possible NWS arms limitation and security policies and, as such, clearly illuminate the possible relationships between these NWS policies and future nuclear proliferation. They fall towards opposite ends of a continuous variable which is a composite of the size of the nuclear weapon inventories of the NWS, the rate of qualitative development of their nuclear weapons, and their strategic doctrines governing the diplomatic and military contingencies in which the deployment, use or threatened use of nuclear weapons is contemplated. These three components of the composite variable are assumed to move together. This composite variable can thus be viewed as a measure of a NWS's overall investment in and political utilization of nuclear weapons.

Finally, given their adversary relationship, Bull assumes that the USA and the USSR will pursue the same doctrine at any given point in time.

These two postures differ in three basic respects: in their major policy prescriptions; in their assumptions about the amount of change in the international security system which is required in order to minimize nuclear proliferation by the year 2000; and in their objective functions.

First, in the High Posture Doctrine, the two major NWS *maximize* the gap between their own nuclear weapon capabilities and those of minor NWS and NNWS by maintaining a large inventory of nuclear weapons, sustaining a rapid rate of qualitative development of nuclear weapons, and relying upon the deployment and threatened use of nuclear weapons in a wide range of diplomatic and military contingencies. Specifically, under the High Posture Doctrine, the major NWS rely on guarantees to use or threaten to use nuclear or conventional force as the major policy instruments with which to deter other NWS from using and NNWS from acquiring nuclear weapons[34b]. Its advocates differ somewhat on the form such threats should take and the range of contingencies in which they should be used.

Second, the High Posture Doctrine projects much of the existing international security system into the year 2000 and rejects substantial changes in existing arms limitation and security policies.

Finally, while designed to minimize nuclear proliferation, among other objectives, the High Posture Doctrine weights maintaining the US-Soviet strategic balance at a high level and the cohesion of US alliances with NATO and Japan above the minimization of nuclear proliferation, should these priority objectives be seen as coming into conflict.

By contrast, in the Low Posture Doctrine, the United States and the Soviet Union *minimize* the gap between their own nuclear weapon capabilities and those of minor NWS and NNWS by maintaining a small inventory of nuclear weapons, constraining the rate of qualitative development of nuclear weapons, and relying upon the deployment and threatened use of nuclear weapons in only a very limited range of diplomatic and military contingencies. In the Low Posture Doctrine, the NWS adopt a comprehensive range of new arms limitations, arms reductions and security obligations in order to deter NNWS from acquiring nuclear weapons[34b].

Moreover, the Low Posture Doctrine, by combining multiple policy instruments into a long-term anti-proliferation strategy, requires major shifts in existing arms limitation and security policies and seeks significant alteration in the international security system by the year 2000.

Finally, while the Low Posture Doctrine is designed to minimize future nuclear proliferation, among other objectives, its advocates disagree as to whether reducing the US-Soviet strategic balance to a low level and minimizing the use or threatened use of nuclear weapons should be weighted more highly than the minimization of nuclear proliferation, should these priority objectives be seen as coming into conflict. Advocates of the modified Low Posture Doctrine, which includes residual positive security guarantees for use against actual or threatened nuclear attack, give maximum weight to the objective of minimizing future nuclear proliferation.

The following two sections summarize the arguments offered on behalf of each of these two postures as efficient anti-proliferation régimes which can to some degree satisfy the policy objectives of the NNWS. In the concluding section it is then argued that the adoption of a modified Low Posture Doctrine by the NWS would best satisfy most of the various policy objectives of the NNWS, thus amplifying the NPT and minimizing future nuclear proliferation.

The High Posture Doctrine

The High Posture Doctrine has been advocated as an anti-proliferation strategy in its extreme form by such analysts as Walter Hahn, Malcolm Hoag and Uri Ra'anan, and by James Schlesinger before he entered the US government in 1969[26d, 29, 36d, 65b, 67]. Recently, in less extreme form, it has been advocated by Alton Frye[68]. This discussion combines their similar arguments concerning how much the NWS, and specifically the United States and the Soviet Union, should invest in and utilize nuclear weapons in their conduct of foreign policy in order to minimize future nuclear proliferation.

In short, the extreme variant of the High Posture Doctrine asserts that there is an inverse correlation between the vertical and the horizontal proliferation of nuclear weapons. In this view, the larger the nuclear weapon inventories of the major nuclear powers, the less likely it is that significant numbers of NNWS will acquire nuclear weapons. Conversely, by reducing their nuclear inventories, the major NWS will encourage the proliferation of nuclear weapons.

According to this variant of the High Posture Doctrine, the USA should and the USSR inescapably will develop and deploy large, sophisticated, flexible nuclear force structures because of the general political competition which exists between them. Such a nuclear force would include the following eight kinds of nuclear capability: (*a*) substantial invulnerable, second-strike, assured destruction capabilities directed against enemy urban-industrial targets as the ultimate reserve to deter nuclear attack upon the NWS's own cities and those of its allies; (*b*) accurate low-yield, air-burst weapons capable of limited, precise strikes against military and economic targets with minimal unintended collateral damage; (*c*) some limited hard-target kill counterforce capabilities both to threaten targets which might jeopardize allies and to deter nuclear attack on its own hard military targets; (*d*) pre-planned targeting options as well as rapid retargeting flexibility built into the strategic force structure; (*e*) tactical nuclear weapons both to deter their use by the enemy and for battlefield use; (*f*) sophisticated command and control capabilities permitting tight political control over all nuclear operations; (*g*) some damage-limiting area defences to protect its own population against small or primitive nuclear attacks by minor NWS; and (*h*) some civil defence to protect its own population against all types of nuclear attack so far as possible[26d, 69a].

In order to maintain and modernize this broad range of nuclear capabilities, advocates of the extreme High Posture Doctrine argue that the USA should and the USSR inevitably will pursue a high rate of qualitative improvements so as to maintain the essential equivalence of the two nuclear force structures. Some advocates, such as Uri Ra'anan, argue, moreover, that the USA should maintain a clearly perceptible nuclear strategic superiority in order to neutralize the Soviet geopolitical advantages and larger general-purposes forces in the overall balance between them[29a]. In fact, with the exception of area defences against ballistic missiles, current US and Soviet forces largely conform to the force requirements of the extreme High Posture Doctrine.

In a less extreme variant recently expressed by Alton Frye, the High Posture Doctrine asserts that there is no causal relationship between the size of the nuclear inventories of the NWS — particularly the USA and the USSR — and the future decisions of NNWS whether or not to acquire an independent nuclear weapon capability. Thus, nuclear arms acquisition and/or arms limitation policies should be determined as a function of competitive US-Soviet interactions irrespective of any putative impact upon horizontal proliferation. To quote Frye on this point:

> Some artful commentators have contended that the on-going strategic competition between the United States and the Soviet Union, described as "vertical proliferation" justifies the decision by India and possibly other states to test nuclear explosives — "horizontal proliferation", as the jargon puts it. This is a canard.
>
> The technological refinements coursing through the Soviet and American arsenals have virtually no bearing in logic or in politics on the inclinations of other states to go nuclear. Soviet and American weapons threaten each other, not the non-nuclear states[20, 68a].

In either variant, however, advocates of the High Posture Doctrine dismiss the argument that the NWS must constrain their own levels of nuclear capabilities in order to constrain future nuclear proliferation.

With respect to the use of nuclear weapons, all advocates of the High Posture Doctrine argue that it is necessary for the NWS to rely upon nuclear weapons in their conduct of foreign policy in order to minimize horizontal proliferation. The extreme High Posture Doctrine asserts that the major anti-proliferation policy instrument of the USA, at least, should be the extension to NNWS of unilateral *positive security guarantees*: that is, undertakings to use or to threaten to use US nuclear forces directly on behalf of certain NNWS against hostile states in a broad range of diplomatic and military contingencies. Of particular importance is the provision of such guarantees to NNWS allies in existing alliances. Such threatened uses of nuclear force by the USA will, it is assumed, be sufficiently credible to deter the USSR, minor NWS and certain NNWS from a range of destabilizing or hostile initiatives which might threaten other NNWS. In some instances the USSR might join the USA in extending cooperative positive security guarantees to certain NNWS. The advocates of the extreme High Posture Doctrine hold, moreover, that at least the United States should remain ambiguous about the range of contingencies in which it will initiate the use of nuclear weapons. Such ambiguity will maximize the credibility of all positive security guarantees, since any explicit listing would tend to exclude by implication the first use of nuclear weapons in other residual contingencies[29]. Despite disclaimers to the contrary, the Schlesinger Doctrine, which advocated limited nuclear response options across the spectrum of the nuclear threat and which still informs US strategic planning, has often been interpreted as expanding the range of diplomatic and military contingencies in which the threatened use of limited nuclear response options would be contemplated [27, 56b].

Recently, Frye has proposed a positive security guarantee according to which, "in the event of a nuclear attack on the territory of a nonnuclear state, the United States and the Soviet Union would undertake to make available to the victim a comparable number and scale of nuclear weapons with which to retaliate"[68b]. Such indirect guarantees would be extended to states outside the existing major alliances of the USA and the USSR. (They would perhaps add to, but would in no way replace, existing direct guarantees which the two powers offer to NNWS allies in NATO and the WTO.) By so extending the benefits of deterrence to states not now members of their respective major alliances, Frye argues that the USA and the USSR would jointly discourage the use of nuclear weapons against NNWS and thus the acquisition of nuclear weapons by NNWS involved in local and regional disputes. However, by providing a NNWS with an appropriate retaliatory capability to use or not as it chose, Frye argues that the two major NWS would be less directly embroiled in local or regional crises than if they were obligated to use their own nuclear weapons directly on behalf of the threatened NNWS. Thus Frye's proposed indirect guarantees, if they could surmount the extraordinary training, logistical, and command, control and communication problems which they raise, might be more credible than direct intervention[68b]. In either variant, however, advocates of the High Posture Doctrine emphasize the primary importance of extending some form of positive security guarantees to protect NNWS from a range of threats to their military security.

In addition to advocating the use of nuclear weapons either directly or indirectly on behalf of threatened NNWS, several proponents of the High Posture Doctrine also propose that at least the USA should upgrade its general-purpose forces so as to be able visibly to fulfil commitments, station forces or actively intervene with such forces on behalf of NNWS. They also support the reliable supply of conventional arms to such NNWS. Through such uses of conventional forces and arms transfers, the High Posture Doctrine seeks to maintain a high threshold at which nuclear weapons would be introduced and at the same time to provide NNWS with some incentives not to acquire independent nuclear weapon capabilities[67a, 69b].

The emphasis in the High Posture Doctrine upon a policy of extended nuclear as well as conventional deterrence across a wide spectrum of risks generates, in turn, continuing requirements for the type of large, sophisticated, flexible nuclear force structures described above. It thus implies that limiting and subsequently reducing nuclear capabilities may be counterproductive. The extreme version of the High Posture Doctrine is compatible with only those arms limitation agreements which permit high quantitative limits on and continued qualitative improvements in nuclear weapons and delivery vehicles. Thus, the quantitative limits and qualitative improvements permitted the USA and the USSR in the 1963 Partial Test Ban Treaty, the 1972 SALT I Interim Agreement, the 1974 Proposed Treaty on the Limitation of Underground Nuclear Weapons Tests, the 1976 associated Treaty on Underground Nuclear Explosions for Peaceful Purposes, and the 1974 Vladivostok Accords are compatible with the extreme version of the High Posture Doctrine. The 1972 and 1974 ABM treaties are not.

The extreme variant of the High Posture Doctrine also implies that constraining the deployments and uses of nuclear weapons through various agreements will be counterproductive. Since it is incompatible with agreements which limit the diplomatic and military contingencies in which nuclear weapons can be deployed and used, the High Posture Doctrine would reject the range of deployment and the use constraints discussed below in connection with the Low Posture Doctrine. Furthermore, at least in theory, deployment and use constraints such as the 1959 Antarctic Treaty, the 1967 Outer Space Treaty and the 1971 Sea-Bed Treaty are incompatible with the extreme version of the High Posture Doctrine. In practice, they are compatible only because the environments and regions affected are of peripheral military concern.

Frye, however, advocates a comprehensive test ban and an international régime for peaceful nuclear explosions as important additional means of constraining horizontal proliferation. He argues, as noted above, that other quantitative and qualitative arms limitation agreements are unrelated to horizontal proliferation and should be pursued, if at all, as a means of stabilizing the bilateral relationship between the USA and the USSR[68b, 70]. Frye also asserts that his proposal is consistent with certain deployment and use constraints, such as nuclear-free zones and certain forms of negative security and no-first-use guarantees[68b]. However, the central element of Frye's proposal is similar to that of the extreme High Posture Doctrine: positive security guarantees play the primary role in minimizing horizontal proliferation and the USA and the USSR are enjoined to maintain sufficiently large inventories of nuclear weapons to be able to execute such guarantees.

In summary, the High Posture Doctrine attempts to maximize the gap between the USA and the USSR on the one hand and all other states, both minor NWS and NNWS, on the other. By maintaining such a gap, the doctrine inevitably perpetuates nuclear arms competition between the USA and the USSR at a high level, thereby permitting each of them to offer its major allies reasonably credible unilateral positive security guarantees. However, the High Posture Doctrine also permits the two nations to dominate lesser states by engaging in cooperative behaviour, either tacitly or explicitly, in a variety of contingencies. These effects of this maximum gap are viewed by advocates of the High Posture Doctrine as useful in satisfying at least some of the policy objectives of some NNWS and thereby minimizing their incentives to acquire independent nuclear weapon capabilities.

Achievement of the military objectives of NNWS

With respect to the military security objectives of various NNWS, advocates argue, first, that the High Posture Doctrine enables either major NWS to deter the other from specific military and diplomatic acts which threaten its own central security interests, including threats to its major allies. The major nuclear power in question can thus achieve objective 1 for its major NNWS allies through providing direct, unilateral positive security guarantees to NNWS threatened with nuclear or conventional attack or nuclear blackmail by the other major NWS.

Threats against allies' military forces or targets, according to the High Posture Doctrine, are to be deterred by the threat of limited strategic or tactical nuclear responses against military and economic targets in the territory of the opposing major power or its military allies. Threats against allies' cities should ultimately be deterred by the threat of strategic attacks upon urban-industrial targets in the territory of the opposing major power or its military allies[26d, 56c, 69c]. More generally, the maintenance by the USA of at least essential equivalence with the nuclear force structure of the USSR is thought by advocates of the High Posture Doctrine to be a necessary base for the positive security guarantees which the USA offers its major allies threatened by the USSR[56d, 71]. Extension of positive security guarantees by the USA and the USSR, respectively, failed in the 1950s and 1960s to prevent France and China from acquiring independent nuclear weapon capabilities. However, US positive security guarantees to its NATO allies and to Japan, and Soviet guarantees to its WTO allies — respectively among the most central security interests of the two major NWS — are hitherto successful examples of this case.

In addition to such direct guarantees, Frye suggests that each major NWS might consider providing NNWS with an appropriate retaliatory capability after it had been attacked by the opposing side.

Given the existing state of mutual deterrence between the United States and the Soviet Union, positive security guarantees extended to major allies by one of them against threats by the other can never be completely credible, despite the centrality of the security interests at stake and the flexible force structures prescribed by the High Posture Doctrine. However, under this doctrine, the NNWS seeking military security against a major NWS is also deterred from acquiring nuclear weapons by the hostile NWS since a High Posture imposes high entry costs on a NNWS which is considering acquiring a nuclear weapon capability in order to deter or retaliate against a major nuclear power. To quote Malcolm Hoag on this point:

> Superpower strategic doctrine, when translated beyond rhetoric into the full spectrum of required capabilities, obviously can affect the *incentives* of allies, and of neutrals under the umbrella of superpower nuclear guarantees to undertake national nuclear programs. The incentives relate to the much-debated credibility issue about nuclear guarantees. The impact of superpower strategic doctrine upon the *costs* of prospective nuclear programs receives less attention, but may be equally important[26d, 65b].

By maintaining a large, flexible and constantly modernized nuclear force structure and a doctrine of extended deterrence, each major NWS makes it difficult for a hostile NNWS to escape its strategic vulnerability even after acquiring an independent nuclear weapon capability. By making it unlikely that a hostile NNWS can achieve objective 1, through the acquisition of nuclear weapons, each of the two major nuclear powers thereby helps to minimize nuclear proliferation. Thus, under the High Posture Doctrine, the combination of high entry costs and reasonably credible unilateral positive security guarantees can minimize future nuclear proliferation both by deterring NNWS from acquiring nuclear weapons and by providing them with an alternative means of achieving this first objective.

Second, the High Posture Doctrine prescribes that the USA and the USSR, whether through unilateral or through tacitly or explicitly cooperative actions, extend some form of positive security guarantees to allied or non-aligned NNWS threatened with nuclear or conventional attack or nuclear blackmail by minor NWS. Unilateral US security guarantees to Japan and Australia against China, the Security Council Resolution 255 of 1968, and whatever tentative and tacit security guarantees the NWS parties to the NPT may have attempted unsuccessfully to extend to India against China are examples of such positive security guarantees designed to achieve objective 2. Clearly, such putative guarantees failed to prevent India from developing a nuclear explosive capability. However, if the two major NWS can extend credible positive security guarantees to NNWS threatened by a minor NWS, they may effectively deter the minor NWS from undertaking a nuclear or conventional attack or nuclear blackmail against the NNWS. To quote James Schlesinger on this point:

> In a showdown with a superpower, a minor nuclear power relying on its own resources will simultaneously be deterred and be subject to disarming . . .
>
> The strongest deterrent to a lesser power's employing its capability is the possibility that a major nuclear power will enter the lists against it[65b].

Under the extreme variant of the High Posture Doctrine, such guarantees can be made credible to NNWS if the major nuclear powers have sufficient counterforce capabilities for controlled nuclear reprisals against military targets in a minor NWS and sufficient damage-limiting capabilities to protect their own populations against fairly primitive attacks by minor NWS[26d]. The Frye proposal attempts to deter nuclear attacks by minor NWS by promising to provide threatened NNWS with an appropriate retaliatory capability. Such guarantees to threatened NNWS — whether direct or indirect — would be most credible if cooperatively executed, either explicitly or tacitly, by the two major nuclear powers against the minor NWS. Indeed, this was the logic justifying Security Council Resolution 255 of 1968. If sufficiently credible, such cooperative guarantees would minimize future nuclear proliferation by providing NNWS with an alternative means of achieving objective 2.

The credibility of such cooperative guarantees in both the extreme and the Frye versions of the High Posture Doctrine would be degraded, however, if both the USA and the USSR were relatively disinterested in one such crisis or if they were both deterred from becoming involved for fear of alienating the minor NWS. Finally, unilateral positive security guarantees — either direct or indirect — offered by one major NWS to the threatened NNWS will not be very credible if the other major NWS supports the minor NWS. In such a situation, given the existing state of mutual deterrence between the two major powers, the guarantor may fear a confrontation and possible escalation into a nuclear exchange. Therefore, the guarantor may be tempted to renege. Thus, in these latter contingencies, positive security guarantees may fail to provide the NNWS with a credible alternative means of achieving objective 2.

Third, the USA and/or the USSR might attempt, according to the extreme High Posture Doctrine, to extend direct positive security guarantees — whether unilateral, tacitly cooperative or explicitly cooperative — to allied or non-aligned NNWS threatened by conventional attack from hostile NNWS. The willingness of the United States to use or threaten to use nuclear force on behalf of Israel against a conventional Arab attack or on behalf of South Korea against a conventional attack by North Korea are possible examples of this case. Given the conventional nature of the threat, the Frye proposal does not apply in this case. A direct guarantee, however, if credible, provides the NNWS with an alternative means of achieving objective 3.

Fourth and finally, the USA and the USSR, whether through unilateral or cooperative actions, might intervene in an incipient local or regional nuclear arms race by providing positive security guarantees to allied or non-aligned NNWS threatened by the prospective acquisition of nuclear weapons by a hostile local or regional NNWS. The willingness of the USSR to use or threaten to use nuclear force on behalf of the Arab states or, under the Frye proposal, to threaten to give an appropriate retaliatory capability to the Arab states in order to deter the acquisition, deployment and use of nuclear weapons by Israel is an example of this case. Such a guarantee, if credible, provides NNWS with an alternative means of achieving objective 4.

However, the extreme High Posture Doctrine is less credible in the cases of objectives 3 and 4 than in the cases of objectives 1 and 2. In cases 1 and 2, each major NWS can offer positive security guarantees to NNWS threatened by hostile NWS — either the other major NWS or a minor NWS — because of its evident willingness to use or threaten to use nuclear weapons to deter nuclear or conventional attack or nuclear blackmail by a hostile NWS. This willingness is less credible in otherwise conventional contingencies. The extreme High Posture Doctrine certainly does not reject the threatened first use of nuclear weapons in cases 3 and 4. Indeed, this position reflects traditional US policy since, with the exceptions of the Treaty for the Prohibition of Nuclear Weapons in Latin America (Treaty of Tlatelolco) and the recent, highly conditioned statement on the non-use of nuclear weapons at the UN Special Session on Disarmament, the USA has sought to retain an ultimate option of first resort to nuclear weapons, particularly tactical nuclear weapons, in otherwise conventional engagements[72]. Thus, their threatened use cannot be ignored as a means of achieving objectives 3 and 4. However, given the increasing legitimacy of the nuclear firebreak, a threat to intervene with controlled nuclear responses, whether unilateral or cooperative, is of declining credibility in these cases, since such local or regional crises or incipient nuclear arms races would otherwise still involve only NNWS.

The Frye proposal explicitly deals with the incredibility of nuclear intervention by the two major NWS in an otherwise conventional local or regional crisis by threatening to provide an appropriate retaliatory capability to a NNWS only after nuclear weapons had been used in such a crisis. Thus, while inapplicable in satisfying objective 3, the provision of such indirect positive security guarantees, if they could be made operational, would better satisfy objective 4 than could the direct positive security guarantees contemplated in the extreme version of the High Posture Doctrine.

Perhaps more importantly, in recognition of the potency of the nuclear firebreak, some advocates of the High Posture doctrine propose upgrading the conventional capabilities of at least the USA to show NNWS that the USA can intervene with conventional forces in order to achieve objectives 3 and 4[26d, 67b].

Like commitments to use nuclear weapons, commitments threatening the use of conventional force in cases 3 and 4 would be most credible if they were cooperatively undertaken — either tacitly or explicitly — by the USA and the USSR. If an important security interest of one major NWS is sufficiently unimportant to the other, what appears to be a unilateral intervention by one would be tacitly acceptable to the other. Alternatively, explicit cooperative commitments might be quite credible if important security interests of both powers converge, enabling them to intervene or interpose forces jointly in such a local or regional crisis[73a]. A joint US-Soviet peacekeeping force in a future Middle East crisis or joint interposition in Southern Africa are conceivable examples. In such instances, the cooperative commitment of conventional forces would minimize future nuclear proliferation by providing NNWS with an alternative means of achieving objectives 3 and 4.

The credibility of the Frye proposal or of cooperative commitments to intervene with conventional forces would be degraded, however, if both major NWS were relatively disinterested and chose to remain uninvolved in such a crisis. Refusal to protect an international pariah state would be a possible example. Furthermore, any unilateral commitment to intervene or to threaten to intervene may not be very credible in local or regional crises or incipient nuclear arms races, such as the Middle East, the Persian Gulf or Southern Africa, in which the two major nuclear powers might be on opposite sides. Such undertakings risk confrontation between the USA and the USSR and possible escalation to a nuclear exchange. Given the state of mutual deterrence existing between them, the guarantor in this instance might be reluctant to engage in nuclear risk-taking and, instead, be tempted to renege. Thus, reliance upon either the Frye proposal or conventional commitments in these latter contingencies may fail to provide the NNWS with a credible alternative means of achieving objectives 3 and 4.

Objectives 2, 3 and 4 all involve countering the actions of local or regional adversaries, whether minor NWS or NNWS, rather than a major NWS adversary as in the case of objective 1. Since the NNWS are not concerned with a major NWS adversary in these three cases, the two major NWS cannot deter them from acquiring independent nuclear weapon capabilities by directly raising the entry costs of such a capability.

However, the advocates of the High Posture Doctrine do emphasize to NNWS involved in local or regional conflicts that the military security risks of acquiring nuclear weapons may run high. First, they argue that nuclear weapons are most likely to be used in what Richard Rosecrance has called 'minor power subgames'. The real risk of nuclear war will arise when primitive, vulnerable nuclear weapon capabilities are deployed in various local or regional conflict situations[36a, 36b]. To quote James Schlesinger on this point:

> If we are to dissuade others from aspiring to nuclear capabilities, what we should stress is that, if weapons spread, they are not likely to be employed against the superpowers. The penalties for proliferation would be paid, not by the United States, or the Soviet Union, but by third countries.

The likelihood that the first nuclear war, if it comes, will originate in and be confined to the underdeveloped world should play a prominent role in any assessment of proliferation's consequences. The tenor of the existing discussion of proliferation has led some people in the underdeveloped countries to conclude that the major powers would be the chief beneficiaries of curtailing the spread. If nuclear spread is to be effectively opposed, it should be made crystal clear just whose security is placed at risk and whose is not[65b].

Second, the advocates of the High Posture Doctrine suggest that acquiring nuclear weapons will increase the vulnerability of a new minor NWS in the nuclear supergame. According to the High Posture Doctrine, the United States and the Soviet Union place maximum reliance upon the extension of some form of positive security guarantees as an anti-proliferation policy instrument. Should either of them intervene in local or regional conflicts, they are likely to utilize nuclear weapons against minor NWS. By contrast, they are at least somewhat less likely to use nuclear weapons in crises otherwise still limited to NNWS. Thus, if a NNWS acquires a nuclear weapon capability, it substantially increases its risk of being targeted either directly by the major NWS' own nuclear arsenals or, under the Frye proposal, by regional adversaries who receive appropriate retaliatory capabilities from the major NWS[36b].

In summarizing the capacity of the High Posture Doctrine to satisfy the military security objectives of the NNWS, it becomes clear that the USA and the USSR each maintains a high level of sophisticated strategic capabilities and provides positive security guarantees primarily in order to protect its major NNWS allies from nuclear or conventional attack or nuclear blackmail by the other power. These are, according to the High Posture Doctrine, respectively the most central security interests of each major NWS. In so doing, the High Posture Doctrine minimizes the incentives of such major NNWS allies to acquire independent nuclear weapon capabilities as a means of achieving objective 1. This is the traditional objective of the anti-proliferation policies of the 1960s. The High Posture Doctrine, despite the limited credibility of its positive security guarantees and conventional commitments in local or regional conflicts, advocates no further arms limitation and security policies by which the two major NWS might attempt to satisfy objectives 2, 3 and 4 of various NNWS. Thus, in a range of adversary relations between neighbouring or regional states in which the USA and the USSR are either disinterested or at odds, the High Posture Doctrine fails to provide NNWS with alternative means of achieving their military security objectives. Rather, the High Posture Doctrine, after warning them of the attendant risks, simply accepts the likelihood that such NNWS may acquire independent nuclear weapon capabilities in order to achieve objectives 2, 3 and 4[26d, 65b].

Achievement of the political objectives of NNWS

With respect to the political prestige objectives of various NNWS, the High

Posture Doctrine does not offer alternative means of achieving them and thus does not effectively operate to minimize proliferation. Rather, by maximizing the two major NWS' investment in and reliance upon nuclear weapon capabilities in the conduct of foreign policy, and thus maximizing their strategic superiority over all other states, the High Posture Doctrine explicitly emphasizes the unique political status to be derived from nuclear weapons. Indeed, the High Posture Doctrine prescribes that the USA and the USSR — despite their own basic conflicts — use their nuclear weapon capabilities to dominate the existing international political hierarchy: to intervene, either unilaterally or cooperatively, in a variety of contingencies involving lesser states, and to exercise a tacit hegemony in the international security system[43, 65a]. By stressing the central utility of nuclear weapons in maintaining the major NWS' superior status, by assuming that the symbolic importance of nuclear weapons is both currently high and cannot be reduced, and by failing to emphasize alternative sources of status, the High Posture Doctrine clearly suggests that nuclear weapons have political utility and can enhance political prestige.

Thus, at best, the High Posture Doctrine fails to provide NNWS with alternative means of achieving enhanced political prestige in relation to any referent group which the NNWS deem important: whether it be the great powers, as in objective 5; an existing alliance, as in objective 6; or particular local groupings, regional groupings, or transregional status cohorts, as in objective 7. It also fails to provide an alternative means of altering the existing distribution of power and status in the international political hierarchy, as in objective 8. At worst, the High Posture Doctrine generates a 'demonstration effect' that nuclear weapons, as the most modern and useful of weapons, enhance political prestige. It thereby provides NNWS with both positive incentives and a useful excuse to acquire nuclear weapons[36c]. The effect of the High Posture Doctrine upon the NNWS' pursuit of each of the four specific political prestige objectives is reviewed below.

First, the High Posture Doctrine asserts that the USA and the USSR should maintain maximum investment in and reliance upon nuclear weapons in order to maintain their supreme great power, that is, major NWS, status. It thereby emphasizes the symbolic political importance of a nuclear weapon capability as a component of great power status[56e]. Thus, NNWS which might plausibly lay claim to great power status are encouraged to acquire nuclear weapons, rather than to build upon some other component of great power status, as a means of achieving objective 5.

Second, the High Posture Doctrine, as described above, places strong emphasis upon providing positive security guarantees to major NNWS allies in formal military alliances as a means of achieving objective 1. However, no proud state in contemporary international politics can comfortably accept a formally inferior and possibly demeaning status as a protectorate of a NWS. Relations between the dominant NWS and the major NNWS allies in formal military alliances have at times been perceived in this fashion. To some degree, NATO mitigated the impact of such status differentiation through the establishment of the Nuclear Planning Group in 1966. However, by emphasizing both the importance of military alliances and the utility of nuclear weapons as a means of enhancing military security, the High Posture Doctrine may, nevertheless, exacerbate such frictions. It thus encourages certain NNWS to acquire nuclear weapons in order to rectify a perceived inferior status within an existing alliance and achieve objective 6. France and China in the 1950s and the 1960s are examples of this case. It also encourages other relevant NNWS, such as India before 1974, to pursue independent nuclear capabilities as a means of asserting their non-aligned status and political independence from existing military alliances.

Third, by emphasizing the central utility of nuclear weapons for the two major NWS as a means of enhancing their own political prestige, the High Posture Doctrine encourages the acquisition of nuclear weapons by any NNWS pursuing second-order power status and hence enhanced political prestige in a particular local grouping, regional grouping or trans-regional status cohort. Such NNWS may see the acquisition of nuclear weapons as an efficient means of establishing or reinforcing leverage and leadership within whatever referent group they deem important, much as the two major NWS rely upon their own nuclear weapon capabilities to exert hegemony in the entire international political hierarchy. Moreover, such NNWS tend to exist in competitive local or regional pairs or groupings. Thus, the acquisition of nuclear weapons by one such NNWS may well induce a local or regional nuclear arms race in which its NNWS competitor subsequently proceeds to acquire nuclear weapons in order to enhance its now diminished political prestige in the same referent group. Since such pairs or groupings of competitive NNWS exist in most major regions of the world, the High Posture Doctrine encourages a fairly large number of NNWS to acquire nuclear weapons as a means of achieving objective 7.

In the previous three cases of objectives 5, 6 and 7, the High Posture Doctrine demonstrates to NNWS the utility of acquiring nuclear weapons in order to raise their own status within the existing international political hierarchy or some subset thereof. Objective 8 challenges the existing hierarchy itself. The High Posture Doctrine, as noted above, places a premium on the two major NWS' maintaining pre-eminent nuclear weapon capabilities in order to dominate the existing international political hierarchy and to intervene, either unilaterally or cooperatively, in a variety of contingencies involving smaller states[65b]. Moreover, the NPT appears to many NNWS as institutionalizing and legitimizing such pre-eminence of the major NWS, since the NPT limits neither their nuclear capabilities nor their reliance upon nuclear weapons in the conduct of foreign policy. A significant number of NNWS may come to view the tacit major NWS hegemony prescribed by the High Posture Doctrine and legitimized by the NPT as unacceptable, particularly to the degree that it threatens military intervention by NWS in their own local and regional affairs. Thus they may be tempted to acquire nuclear weapons in order to erode such hegemony and alter the existing international political hierarchy in the direction of greater equality among states. Thus, by encouraging NWS hegemony in the international security system, the High Posture Doctrine also encourages NNWS to acquire nuclear weapons as a means of achieving objective 8[27].

In considering these effects of the High Posture Doctrine upon the political prestige objectives of NNWS, some advocates of the doctrine warn that a primitive nuclear weapon capability will not buy much political prestige, just as they warn of the high military risks attendant upon acquiring a primitive nuclear weapon capability. They argue that it is the large size, strategic flexibility and sophisticated quality of the major NWS' nuclear weapon capabilities which provide them with their unique political status and that a more modest capability will not bring commensurate prestige. Nuclear weapons, in this view, are not a 'great equalizer' in providing states with political prestige any more than they are in providing states with military security[26d, 65b].

Despite this qualification, however, other advocates of the High Posture Doctrine admit that the doctrine may encourage nuclear proliferation by emphasizing the general symbolic importance of nuclear weapons and by demonstrating in each of the above cases that nuclear weapons can enhance political prestige. To quote Malcolm Hoag on this effect:

> This position . . . on . . . superpower "vertical proliferation" would clearly not maximize the demonstration effect against "horizontal proliferation". On the contrary, others would claim that the superpowers are setting an example of an arms race rather than arms-control restraint. To them this position would supply a prominent rationale for not signing the NPT . . . [T]his demonstration effect must be viewed as a serious, if virtually inevitable, defect of this position[36d].

Thus, the High Posture Doctrine fails to provide NNWS with alternative means of achieving objectives 5 to 8. Moreover, through the demonstration effect, it actually encourages NNWS to acquire nuclear weapons in order to achieve their political prestige objectives.

The Low Posture Doctrine

The Low Posture Doctrine has been advocated as an anti-proliferation strategy in its extreme form by such analysts as Richard Falk, and Hedley Bull in his recent writing, and in modified form by Leonard Beaton, John Maddox, Max Singer and Ian Smart[9b, 34, 42, 60, 65a, 66, 73b, 74]. As in the discussion of the High Posture Doctrine, their similar arguments concerning how much the NWS, and particularly the USA and the USSR, should invest in and utilize nuclear weapons in the conduct of foreign policy in order to minimize the proliferation of nuclear weapons, can be combined.

In short, the Low Posture Doctrine assumes that there is a positive correlation between the vertical and the horizontal proliferation of nuclear weapons: the smaller the NWS' inventories of nuclear weapons, the less likely it is that significant numbers of NNWS will acquire nuclear weapons. Conversely, by enlarging their nuclear inventories, the NWS will encourage the proliferation of nuclear weapons.

With respect to nuclear force structures, all advocates of the Low Posture Doctrine argue that the NWS, and particularly the USA and the USSR, should each develop and acquire only a relatively small inventory of invulnerable, second-strike weapons. Such stable, mutual assured destruction capabilities would serve as a deterrent designed solely to deter or retaliate against nuclear attacks by hostile NWS in order to maintain or restore a non-nuclear *status quo*[73b, 75]. Such assured destruction capabilities have traditionally been designed to carry out massive attacks on the enemy's urban-industrial targets. However, massive retaliation is not a necessary targeting requirement of the Low Posture Doctrine; both Richard Falk and Herman Kahn have advocated limited proportional responses in strict retaliation for a nuclear attack by any NWS adversary as an appropriate alternative targeting strategy[65c, 73b].

Assuming that the sole function of nuclear weapons is to deter or retaliate against nuclear attacks, the Low Posture Doctrine explicitly opposes massive hard-target kill counterforce capabilities, defensive damage-limiting capabilities, and tactical nuclear war-fighting capabilities. It also emphasizes the firebreak between nuclear and conventional weapons by opposing the acquisition of such potentially ambiguous weapon systems as dual-capable delivery vehicles and mini-nukes. Furthermore, some advocates of the Low Posture Doctrine also seek to raise the nuclear threshold by upgrading conventional military capabilities. In sum, any military conflicts other than retaliation against nuclear attacks are to be waged without resort to nuclear weapons.

In order to maintain such a force structure, only a relatively low rate of qualitative development is required. All advocates of the Low Posture Doctrine argue that the NWS should pursue only those improvements in their nuclear weapon systems which are necessary to maintain a credible assured destruction capability. Any qualitative developments which might destabilize their mutual assured destruction capabilities or provide counterforce or damage-limiting capabilities should be avoided.

In order to meet only these limited nuclear force requirements, the Low Posture Doctrine stipulates that the NWS, and particularly the USA and the USSR, should constrain and subsequently decrease their existing strategic capabilities through a combination of quantitative and qualitative arms limitation and reduction agreements. Quantitative reductions in particular are required since, as Paul Doty has argued, qualitative agreements can constrain but never completely freeze qualitative improvements in strategic force structures of a given size. Thus, only quantitative reductions can hold constant and subsequently reduce overall strategic capabilities[76]. Advocates of the Low Posture Doctrine support such quantitative arms limitation and disarmament agreements as: reduced ceilings on strategic offensive nuclear weapons and delivery vehicles in follow-on agreements to the Vladivostok Accords and SALT II; a cut-off of production of fissile material for military purposes; limitations on attack submarines; a prohibition of anti-ballistic missiles; and reductions of tactical nuclear weapons, specifically eliminating those appropriate for war-fighting rather than for deterrence of nuclear attack[27, 60b]. Advocates of the Low Posture Doctrine also support such qualitative agreements as: a Comprehensive Test Ban; limitations on missile test firings; restrictions on various attributes of strategic offensive nuclear weapons and delivery vehicles, such as their size, range and number of warheads; and restrictions on various attributes of anti-submarine warfare[76,77].

The Low Posture Doctrine justifies a limited nuclear force structure and a low rate of qualitative development on the ground that the NWS should rely on the actual or threatened use of nuclear force solely to deter or retaliate against nuclear attacks. To achieve this restrictive role for nuclear weapons, all advocates of the Low Posture Doctrine stipulate that the NWS, and particularly the USA and the USSR, should agree to undertake a range of constraints on the permissible deployments and uses of nuclear weapons: constraints which, in principle, the extreme High Posture Doctrine explicitly rejects. The form of agreement on such constraints might be tacit understandings or formal, binding legal obligations. Participation in them might be unilateral, bilateral or multilateral. Finally, the substantive coverage of the undertaking might be made subject to certain conditions or be absolutely categorical. Each of these characteristics falls on a continuum and, over time, a constraint on deployment or use might become more legally binding, include more participants, and/or become more comprehensive in its substantive coverage.

Whatever the form of the agreement, however, advocates of such deployment and use constraints argue that the permissible limits in any such constraint be as explicit, conspicuous and unambiguous as possible. In this way the constraint will clarify both for other NWS and for NNWS the precise, limited, diplomatic and military contingencies in which nuclear weapons will be deployed and/or used. In contrast to the value which the High Posture Doctrine places upon strategic flexibility, the Low Posture Doctrine seeks to minimize ambiguity and discretion in the deployment and use of nuclear weapons. According to the Low Posture Doctrine, such uncertainties tend to expand the range of diplomatic and military contingencies in which recourse to nuclear weapons would be contemplated by the bureaucratic actors involved, and thus increase the probability of nuclear weapons being utilized in war[73b]. Conversely, minimal uncertainty decreases the probability of nuclear weapons being used in various contingencies.

The Low Posture Doctrine rests on the assumption that one major incentive for NNWS to acquire nuclear weapons is to protect themselves against the nuclear weapons of existing NWS under present and future governments since, without nuclear weapons, a NNWS is at an absolute disadvantage against a NWS should a confrontation erupt between them[78]. Therefore, explicit undertakings by the NWS to constrain the deployment and use of their nuclear capabilities can reduce the existing incentives for NNWS to acquire nuclear weapons in order to deter, defend or retaliate against the use or threatened use of nuclear weapons by the NWS. Furthermore, to the degree that such undertakings will, over time, strengthen international behavioural norms proscribing the development, deployment and use of nuclear weapons, they will inhibit NNWS engaged in local or regional conflict situations with other NNWS from acquiring and subsequently using nuclear weapons.

In order to minimize future nuclear proliferation, advocates of the Low Posture Doctrine thus support one or more of four major types of constraint upon the permissible deployments and uses of nuclear weapons: non-deployment zones; nuclear-free zones; negative security guarantees; and no-first-use agreements. Furthermore, they either support a restricted role for or reject positive security guarantees. Since these proposals substantially diverge from the existing doctrines of the NWS, they are described in some detail below.

One type of constraint, solely limiting the deployment of nuclear weapons, is a non-deployment or nuclear disengagement zone in which NWS external to a particular zone undertake to remove nuclear weapons from the zone and, perhaps, from mobile stations such as naval vessels which regularly operate near the zone. Such undertakings may be tacit or formal; unilateral, bilateral or multilateral; and conditional or categorical. The NWS subject to the constraint could make the undertaking verifiable by accepting some form of permanent observation or verification by challenge. Such a deployment constraint is not necessarily coupled with any constraints on the use of nuclear weapons in the zone; thus, nuclear weapons deployed outside the zone might in future be utilized against targets within the zone. However, an agreement to remove nuclear weapons, particularly tactical nuclear weapons, from a specific zone does prevent the regular inclusion of such stationed nuclear weapons in contingency planning. To the degree that nuclear weapons are disengaged from the zone, military planning for the zone can be decoupled from immediate reliance on accessible nuclear weapons. Thus, such a deployment constraint at a minimum reduces the probability of nuclear weapons being used early in local conflicts. Furthermore, several such deployment constraints would collectively reduce the locational proliferation of nuclear weapons, and thereby reduce at least to some degree the force requirements for certain types of tactical nuclear weapons.

A second type of agreement constraining both the deployment and the use of nuclear weapons in a particular geographic zone is a nuclear-free zone. Such agreements combine undertakings by NNWS in the affected zone not to acquire independent nuclear weapon capabilities nor to permit nuclear weapons to be stationed on their territories with undertakings by

NWS external to the zone not to provide NNWS in the zone with independent nuclear weapon capabilities nor deploy nor use their own nuclear weapons in the affected zone. In effect, a nuclear-free zone combines a zonal non-acquisition and non-dissemination agreement, a non-deployment zone and a zonal negative security guarantee. Such undertakings are likely to be formal and multilateral, including both NWS and NNWS parties, although they can be either conditional or categorical. They also are likely to include some regional control and verification capabilities. The Treaty of Tlatelolco is the existing example of such an agreement.

Nuclear-free zones impose quite pervasive constraints on the deployment and use of nuclear weapons in the affected zone. Thus, they also impose substantial political preconditions, since a region must have a fairly high degree of political cooperation even to contemplate such an undertaking. Indeed, it may be necessary to resolve important regional political disputes before any such nuclear-free zone can be established. However, if this prerequisite political stability could be established, the pervasive constraints which nuclear-free zones impose would make them an effective means of minimizing future nuclear proliferation in the zone[60c]. A regional zone-of-peace agreement, such as that proposed for the Indian Ocean, which incorporates these constraints on nuclear weapons within some broader demilitarization arrangement, would have similarly pervasive effects.

Both a non-deployment zone and a nuclear-free zone prohibit NWS external to the zone from deploying their own nuclear weapons in the affected zone. Of course the NWS could easily reverse such undertakings, since their nuclear weapon capabilities are still in existence elsewhere. Yet, at any particular time, constraints on deployment involve the absence of certain physical capabilities in particular locations. They are thus subject to verification either by continuous observation or by challenge.

By contrast, simple constraints on particular uses of nuclear weapons not coupled with further constraints on their deployment or possession are purely declaratory statements of intention by the NWS in question. For this reason, they have often been faulted — particularly by Western states — for their intrinsic unverifiability and unenforceability. However, most advocates of the Low Posture Doctrine argue that constraints on the permissible uses of nuclear weapons are important means of reducing the probability of nuclear weapons being utilized in war, and therefore of minimizing the future proliferation of nuclear weapons.

Thus, a third general type of constraint on the deployment and uses of nuclear weapons is the *negative security guarantee*: that is, an undertaking by one, some or all NWS not to use nuclear weapons against certain NNWS. Such guarantees may be tacit or formal; unilateral, bilateral or multilateral; and conditional or categorical. As noted before, they are intrinsically unverifiable and unenforceable. However, certain negative security guarantees may strongly imply that NWS redeploy those stationed nuclear weapons with characteristics of range, yield, accuracy and/or mobility appropriate only for the prohibited use. In this way, certain negative security guarantees could be tied to the absence of certain physical capabilities in particular locations and thus be made somewhat verifiable.

Advocates of the Low Posture Doctrine have proposed a variety of negative security guarantees, as did various NNWS during negotiations on the NPT. One common variant is a negative security guarantee extended to NNWS parties to the NPT, thus providing these states with a counter-discriminatory benefit in return for their undertaking not to acquire independent nuclear weapon capabilities[9b]. While the NWS parties to the NPT have refused to offer such a negative security guarantee, four NWS have extended such a guarantee to parties to the Treaty of Tlatelolco in Additional Protocol II of that treaty.

A second common variant is a negative security guarantee extended to some or all NNWS which do not involve themselves in some way in the military alliances or military actions of NWS. Such a condition may take the form of extending negative security guarantees to all non-aligned NNWS. Alternatively, it may take the form of extending negative security guarantees to all NNWS which do not permit NWS to deploy nuclear weapons on their territories. This form is represented by the Kosygin Proposal of 1966. Finally, such a condition may take the form of extending negative security guarantees to certain NNWS provided that they are not then engaged in armed combat in concert with a NWS. This form is represented by the recent US statement on the non-use of nuclear weapons at the UN Special Session on Disarmament[38,44, 79, 80a, 308].

A third common and categorical variant is a negative security guarantee extended to all NNWS irrespective of their position *vis-à-vis* the NPT or their military involvement with the various NWS[27, 34, 66]. Whatever the variant, all negative security guarantees are defended on the ground that it is only the nuclear weapon capabilities of existing NWS which can currently threaten NNWS with nuclear attack or nuclear blackmail. Thus, any prohibition against such attacks is designed both to enhance the military security of affected NNWS and to provide them with an incentive to abstain from acquiring independent nuclear weapon capabilities.

The fourth general type of constraint on the deployment and use of nuclear weapons is an undertaking by NWS not to initiate the use of nuclear weapons against other NWS. Such an undertaking of no-first-use may be tacit or formal; and unilateral, bilateral, or multilateral. It is typically a categorical renunciation of the option to initiate a nuclear exchange. It does not require redeployment, cessation of production and stockpiling, or subsequent reduction of stationed nuclear weapons since their retaliatory use is explicitly permitted under such a no-first-use constraint. However, it may strongly imply redeployment, cessation of production and stockpiling, and possibly reductions of those weapon systems peculiarly appropriate for first use and/or peculiarly inappropriate for retaliation[81].

Some advocates of the Low Posture Doctrine such as Richard Falk, Max Singer and recently Hedley Bull have strongly supported a universal prohibition against the first use of nuclear weapons against other NWS [27, 40, 73b, 81]. They argue that such a no-first-use undertaking, in conjunction with negative security guarantees to NNWS, would explicitly restrict the function of nuclear weapons to that of deterring or retaliating against a nuclear attack in conflicts involving other NWS. Such an undertaking thus reduces "the immediacy of present feelings that nuclear weapons are a vital instrumentality" of foreign policy for NWS; reduces discretionary initial resort to nuclear weapons in crises involving NWS; prevents NWS from treating nuclear weapons as efficient technological adjuncts to conventional capabilities and thereby strengthens the firebreak between conventional and nuclear weapons; reduces the incentives of the NWS adversary to pre-empt; and thus reduces the probability of nuclear war[27, 73b]. In this way, a no-first-use undertaking forces relationships among NWS to revert to a 'pre-nuclear' basis with respect to the pursuit of their military security and political prestige objectives[65c, 66, 73b].

Advocates of a no-first-use undertaking admit that such an undertaking may, by decreasing the risk of escalation to a nuclear conflict, somewhat increase the probability of conventional attack upon a NWS or its close military allies by a NWS adversary. Thus, they typically support the maintenance of sufficient general-purpose forces to deter or retaliate against conventional attacks by other NWS[40b, 73b, 81].

Furthermore, advocates of the Low Posture Doctrine admit that the value of purely declaratory negative security guarantees and no-first-use undertakings may be slight at their moment of issuance. However, they argue that such constraints on use will build upon the tradition of non-use of nuclear weapons which has been evolving since 1945, despite frequent threats of their use in various contingencies. Moreover, they argue that the value of explicit constraints upon the permissible uses of nuclear weapons will increase the longer they are observed. Thus, such constraints may, over time, grow into compelling international behavioural norms which impose explicit and substantial limits upon the behaviour of a given NWS and thereby mould the expectations of both NNWS and other NWS. If, for example, it is national policy — whether a tacit or, more compellingly, a legally binding policy — not to use nuclear weapons in certain contingencies against either NNWS or other NWS, involved bureaucratic actors in a given NWS will be less likely and less able to plan to do so. Similarly, if NNWS are assured continually that nuclear weapons will not be used against them — either under certain conditions or categorically — they may come to feel less threatened. NNWS may thus be less tempted to acquire independent nuclear weapon capabilities, particularly since the exercise of their nuclear weapon options would deprive them of the protection of an existing negative security guarantee. Furthermore, if a no-first-use undertaking is in force, the military and political utility of nuclear weapons will appear to decline and NNWS will be less likely to acquire them in order to enhance their military security and/or political prestige[82]. Finally, if other NWS are assured continually that they will not suffer an initial nuclear attack, they also may come to feel less threatened. They may then be more willing to constrain their own uses of nuclear weapons, and thus more willing to redeploy, discontinue producing and stockpiling, and reduce certain nuclear weapon systems in their inventories.

Even if future nuclear proliferation were to occur, the existence of permanent constraints upon the use of nuclear weapons would make the advent of a new NWS safer for the international security system than if such constraints were not in force. A new NWS would be immediately expected to conform with such use constraints as nuclear-free zones, negative security guarantees, and no-first-use undertakings. If it failed to conform, its intentions would be inherently suspect and pressure could be brought to bear on it by other NWS at a time when its new nuclear capabilities were likely to be most primitive and hence most vulnerable[40b].

Finally, advocates of the Low Posture Doctrine, such as Richard Falk, while admitting that such guarantees of contingent or categorical non-use of nuclear weapons are intrinsically unverifiable and unenforceable, argue that they need not be absolutely reliable to be worthy of support. Even a purely declaratory undertaking will raise the marginal costs of engaging in prohibited actions in the calculations of relevant decision-makers. Thus, the creation of such inhibitions, even if they are not absolute, is a valuable policy achievement. To quote Falk on this point in reference to his proposal for a no-first-use undertaking:

> The status of a weapon does appear to have some bearing upon the decision to use it. The argument in favor of a no-first-use position need not sustain the burden of arguing that it will prevent an initiating recourse to nuclear weapons under all circumstances. It is sufficient to show the creation of a significant inhibition in some crucial situations. No rule that is worth formulating anticipates absolute compliance[40a].

Advocates of the Low Posture Doctrine often defend their emphasis upon contingent or categorical constraints on the use of nuclear weapons by citing the precedent of the Geneva Protocol of 1925 as a use constraint of proven effectiveness. The Geneva Protocol established a norm of prohibition against the first use of chemical and biological weapons in war. Since 1925, the Protocol has changed attitudes, expectations and behaviour in the international security system concerning chemical and biological weapons by denying legitimacy to and increasingly stigmatizing their use. Chemical and biological weapons have thus come to be viewed with distaste and revulsion, and while states may retain them, they do not flaunt them or treat them as a 'normal' means of waging war[40b]. The only legitimate use of chemical and, until 1971, biological weapons has been generally thought to be that of deterring or retaliating against their first use by another state. Those states which have introduced chemical weapons in war, such as Italy's campaign in Ethiopia in 1936 and, most recently, the US involvement in Indo-China in the 1960s and early 1970s, have incurred opprobrium and loss of political prestige as a result.

Furthermore, the first use of chemical weapons in warfare by the USA in Indo-China encouraged various efforts at the UN to strengthen the

Geneva Protocol: first, by encouraging universal adherence to it; second, by interpreting its coverage as comprehensive and thereby including irritants, such as tear-gas, and anti-plant agents, such as herbicides and defoliants; and, third, by removing reservations limiting its applicability only to parties[80, 83]. Thus, over time, the Geneva Protocol has substantially changed general attitudes, expectations and behaviour concerning chemical and biological weapons. By rendering such weapons politically illegitimate, the Geneva Protocol has minimized the number of states which explicitly rely upon them to achieve various military security and political prestige objectives.

Moreover, since 1966, the Geneva Protocol has also reinforced further efforts at the UN and the Conference of the Committee on Disarmament (CCD) to achieve limitation and disarmament of the biological and chemical agents used in warfare. The Biological Weapons Convention, endorsed by the UN General Assembly in 1971, achieved complete disarmament in biological and toxin weapons. Draft conventions to prohibit the development, production, acquisition, stockpiling and retention of at least some of the most lethal chemical warfare agents are currently under consideration at the CCD[83]. Thus, to some degree the Geneva Protocol has facilitated subsequent arms limitation and disarmament of chemical and biological weapons.

By analogy, proponents of the Low Posture Doctrine argue that constraints on the permissible deployments and uses of nuclear weapons will, like the Geneva Protocol, minimize future nuclear proliferation, denuclearize international politics and, over time, rationalize and facilitate nuclear arms limitation and disarmament[73b]. In the case of nuclear weapons, as in chemical and biological weapons, the process which achieves these two objectives is likely to be an iterative one. A declining trend may emerge by which certain constraints upon the deployment and use of nuclear weapons generate reductions in overall strategic capabilities which, in turn, permit only limited deployments and uses of the remaining inventories. Limited missions will thus ultimately require only relatively small inventories of invulnerable second-strike weapons in a small number of NWS.

As described above, proponents of the Low Posture Doctrine advocate constraints on the permissible deployments and uses of nuclear weapons in order to minimize their role in international politics. Thus, they all reject the High Posture Doctrine's reliance upon unilateral or cooperative positive security guarantees as the primary, generally applicable, anti-proliferation policy instrument which NWS should use in support of NNWS in a broad variety of diplomatic, conventional military or nuclear contingencies.

Rather, proponents of the modified Low Posture Doctrine view positive security guarantees as exceptional and residual anti-proliferation policy instruments. They are exceptional in that such guarantees should be relied upon in only the small number of cases involving actual or threatened nuclear attack against specific NNWS. They are residual in that such guarantees should be relied upon only in those cases in which negative security guarantees, other constraints on deployments and uses of nuclear weapons, diplomatic efforts to achieve peaceful resolution of disputes, interposition of international peacekeeping capabilities, the transfer of conventional arms or conventional military intervention are clearly inappropriate or ineffectual. Moreover, proponents of the extreme Low Posture Doctrine view positive security guarantees as entirely unacceptable. In sum, the Low Posture Doctrine differs from the High Posture Doctrine in that positive security guarantees—if used at all—are an anti-proliferation policy instrument of last rather than first resort.

The Low Posture Doctrine rejects the extension of positive security guarantees to NNWS against an actual or threatened conventional attack by either hostile NNWS or a hostile NWS on several grounds. First, such a guarantee would clearly abridge either negative security guarantees or a no-first-use undertaking. Moreover, it creates a system-wide precedent legitimizing the first use of nuclear weapons in all conventional engagements against NNWS as well as NWS[66, 73b]. More generally, by making the recipient NNWS dependent upon the guarantor's nuclear weapons to achieve its military security objectives, such positive security guarantees enhance the general utility of nuclear weapons in foreign policy both for the recipient and for the guarantor. Thus, they perpetuate rather than minimize the existing reliance upon nuclear weapons in international politics[27].

This position necessarily precludes NWS from using or threatening to use nuclear weapons to defend a close military ally even when there is no effective alternative defence and a conventional military 'defeat' of the NNWS ally is otherwise inevitable[40a]. Proponents of the Low Posture Doctrine argue that this situation should, if possible, be obviated through the peaceful resolution of disputes, the interposition of peacekeeping capabilities, the provision of military assistance, and the upgrading of conventional forces, both indigenous and allied. However, *in extremis*, they conclude that the cost of the conventional military defeats which would be incurred as a result of forgoing positive security guarantees against conventional attack in specific instances would be less than the benefits to be gained from denuclearizing the international security system through disavowing the first use of nuclear weapons in all conventional conflicts.

This argument extends to such cases as unilateral US positive security guarantees to South Korea in the event of an actual or threatened conventional attack by North Korea, assisted by the USSR or China, and to West European NATO allies in the event of a conventional attack or threatened attack by the WTO. NATO can be viewed as perhaps the most prominent case in which a NWS currently offers a positive security guarantee to NNWS against actual or threatened conventional attack. NATO's security is of paramount interest to the USA, and NATO currently confronts numerically superior conventional forces in Central Europe. However, even with respect to this 'best case', proponents of the Low Posture Doctrine such as Richard Falk, Max Singer and, recently, Hedley Bull argue explicitly that the United States should not offer its NATO NNWS allies a positive security guarantee against actual or threatened conventional attack by the WTO[26, 66, 73b]. Bull bases his position in part upon the observation that the credibility of such guarantees is inevitably eroding at present and thus they are increasingly ineffectual means of deterring conventional attack. Falk argues further that the probability of any such attack by the WTO is so slight as to be virtually incredible, and that, were such a conventional attack to materialize, NATO should be able to defend itself by conventional means. Even in the more likely contingencies of escalating crises or wars 'transferred' to Central Europe from non-European theatres, these strategists would counsel against the first use of nuclear weapons to counter conventional attacks.

However, both Bull and Falk also admit that there is a trade-off between US positive security guarantees to NATO against actual or threatened conventional attack by the WTO and the probability of future nuclear proliferation by the affected NNWS in Western Europe. Should the USA adopt a no-first-use doctrine and cease to provide positive security guarantees against actual or threatened conventional attacks, some West European NNWS might be more likely to acquire either national or some form of joint nuclear weapon capabilities in order to deter or defend against such attacks themselves. Although advocates of the Low Posture Doctrine oppose further nuclear proliferation and believe such independent European nuclear weapon capabilities to be unnecessary, Falk at least argues that in this instance, nuclear proliferation would be preferable to a continuing US positive security guarantee. In his view, national or joint European nuclear weapon capabilities would be inherently limited to deterring massive conventional or nuclear attacks upon Western Europe. They would thus be less subject than a US positive security guarantee to expansion into a general system-wide precedent for initiating the use of nuclear weapons in conventional conflicts[9b, 40a, 57, 65c, 66, 73b]. To this degree, proponents of the Low Posture Doctrine value the denuclearization of NWS' foreign policy behaviour above further nuclear proliferation to at least one class of NNWS facing a particular conventional threat.

As noted above, proponents of the Low Posture Doctrine differ on the question of whether NWS should extend either unilateral or cooperative positive security guarantees to NNWS against actual or threatened nuclear attack by hostile NWS. Proponents of the modified Low Posture Doctrine — Beaton, Maddox, Singer and Smart — view such guarantees as an inextricable component of a comprehensive anti-proliferation régime. In their view, the NWS must offer positive security guarantees against nuclear attack to at least some NNWS as an adjunct to some form of negative security guarantee. Beaton and Singer take the broadest position that the NWS should undertake both to refrain from attacking or threatening any NNWS with nuclear weapons and to come to the assistance of any NNWS so attacked or threatened by another NWS[34a, 66]. Smart, more restrictively, supports positive security guarantees to NNWS party to the NPT coupled with an undertaking by NWS not to attack any NNWS party to the NPT which neither has nuclear weapons on its territory nor belongs to NATO or to the WTO[9b].

Whatever the scope of such positive and negative security guarantees, Beaton and Smart argue explicitly that all NWS should undertake to provide them in a multilateral agreement among themselves. Such a multilateral agreement among the NWS should be broader and more binding than Security Council Resolution 255 which is widely viewed as being ineffectual[60b]. Under such an agreement, any nuclear attack on a NNWS protected by the negative security guarantee would constitute an explicit breach by one NWS of its agreement with all other NWS. Thus, it would automatically provoke a strong political reaction and bring the positive security guarantees into play[9b, 34d].

The positive security guarantees thus provided should, according to Beaton, be both joint and several, involving the NWS as collective as well as unilateral guarantors[34d]. Should one NWS attack or threaten to attack a NNWS with nuclear weapons, thus breaching its negative security guarantee, all other NWS might engage in cooperative retaliation on behalf of the guaranteed NNWS, including a nuclear response if necessary. Alternatively, should such an actual or threatened attack occur, one NWS could fulfil its unilateral obligations to the guaranteed NNWS under attack without any cooperative action by other NWS. Such a guarantee could thus convey to the close military allies of one NWS the sense of long-term commitment which would minimize their political isolation and hence their incentive to acquire independent nuclear capabilities.

This sense of efficacious unilateral positive security guarantees can be facilitated through what Leonard Beaton has termed the processes of 'cognizance' and 'commitment' within existing alliances[42b]. By cognizance, Beaton means that the most important military powers systematically share with smaller powers knowledge about the military forces and security policies which affect these smaller powers. It involves no transfer of control over such forces and policies. The Nuclear Planning Group in NATO is, in Beaton's view, an important vehicle of cognizance which permits those NNWS whose security depends in some contingencies upon US positive security guarantees to participate in planning for such contingencies. Access to information, participation in planning, and the ability to hold the NWS guarantor to account for its force levels and strategic doctrines are means which can reduce the status differential between the NWS and their NNWS allies in a military alliance. Even more profound would be the 'commitment' of nuclear forces to a common international security organization. Again, the committed and assigned forces in NATO serve as an example. While the NWS allies in NATO retain ultimate national control and ownership of their nuclear forces, so long as they are committed to NATO they are subject to common planning and are available for common purposes. Thus the processes of cognizance and commitment strengthen the credibility of the unilateral positive security guarantees available to NNWS allies and thereby minimize future nuclear proliferation.

Leonard Beaton applies these principles of cognizance and commitment to what is perhaps the ultimate expression of joint and several positive security guarantees: the international security force as proposed by Beaton in 1972 in *The Reform of Power*[42c]. Under this proposal, all NWS are to commit and eventually assign their entire strategic nuclear capabilities to an international security organization. Ultimate control of the various national nuclear forces would remain with the individual NWS; upon a formal, public declaration, a NWS could withdraw its nuclear forces from the international organization. Furthermore, any threatened use of nuclear weapons by a participating NWS would immediately dissolve the international organization. Thus, individual NWS could continue to protect their own security interests and those of particular NNWS to which they chose to extend positive security guarantees. As long as the international security force were in being, however, operational control of the assigned nuclear capabilities would be under the management of the international security organization. The purpose of this organization would be the maintenance of an invulnerable nuclear force capable of deterring any NNWS from developing nuclear capabilities which could threaten the force. Thus, NWS, through the international security force to which they had assigned their nuclear forces, could collectively represent and protect the security interests of NNWS both against surprise attack by one of the participating NWS and against regional NNWS adversaries which might subsequently acquire nuclear weapons.

Not all proponents of the Low Posture Doctrine support the extension of unilateral or cooperative positive security guarantees to NNWS facing actual or threatened nuclear attack by hostile NWS. In particular, proponents of the extreme variant of the Low Posture Doctrine — Bull and Falk — prefer to rely solely upon negative security guarantees to prevent actual or threatened nuclear attacks upon NNWS by hostile NWS. They oppose positive security guarantees against nuclear attacks on two grounds. First, they argue that such guarantees are of declining credibility and thus are ineffectual means of deterring nuclear attack on NNWS. Second, they argue that all positive security guarantees exploit, rather than minimize, the role of nuclear weapons in international politics. In their view, the sole appropriate use of nuclear weapons for a NWS is to deter or retaliate against a nuclear attack upon itself[27].

In sum, by limiting and reducing nuclear capabilities, constraining permissible deployments and uses of nuclear weapons, and minimizing or, in the extreme, rejecting reliance upon positive security guarantees, the Low Posture Doctrine seeks gradually to de-emphasize and ultimately to delegitimize nuclear weapons. Through the adoption of such policies, NWS would find nuclear weapons of decreasing military and political utility. No longer could NWS rely upon them in a range of military contingencies or claim great power status by virtue of possessing them. The only legitimate role of nuclear weapons for a NWS would ultimately be the separate, residual and peripheral one of deterring or retaliating against nuclear attack upon itself and, perhaps, upon certain NNWS. Thus, the international security system would become gradually denuclearized and in such a denuclearized system, any state utilizing nuclear weapons for any purpose other than deterring or retaliating against nuclear attack would invite opprobrium, loss of political prestige and most importantly, military sanctions, possibly including nuclear sanctions. Max Singer has termed such a denuclearized international security system a "non-utopian, non-nuclear future world"[66]. This is not a utopian world in which the NWS would soon relinquish all their nuclear weapons. Rather, it is an international security system in which nuclear weapons would be reduced, the remaining inventories segregated from the ordinary course of international politics, and the NWS possessing them subjected to special obligations and special constraints.

Given this objective of denuclearizing international politics, the Low Posture Doctrine is compatible with any arms limitation agreement and any deployment or use constraint which imposes some physical or doctrinal restrictions upon the NWS' acquisition and political utilization of nuclear weapons. It is, conversely, incompatible with any arms limitation agreement or doctrine which either enhances or fails to restrict the nuclear inventories or the strategic flexibility of NWS. Thus the ABM Treaties of 1972 and 1974 and the four post-war treaties constraining the deployment and use of nuclear weapons by NWS — the Antarctic Treaty of 1959; the Outer Space Treaty of 1967; the Treaty of Tlatelolco of 1967; and the Sea-Bed Treaty of 1971 — are compatible with the Low Posture Doctrine in that they impose, at least to some degree, actual or potential restrictions upon the nuclear capabilities and doctrines of the NWS. Conversely, by either enhancing or failing to restrict the nuclear capabilities of NWS, the Partial Test Ban Treaty of 1963, the SALT I Interim Agreement of 1972, the Proposed Treaty on the Limitation of Underground Nuclear Weapons Tests of 1974, the associated Treaty on Underground Nuclear Explosions for Peaceful Purposes of 1976, and the Vladivostok Accords of 1974 are not compatible with the Low Posture Doctrine.

The Low Posture Doctrine seeks to minimize the military security and political prestige gap between the NWS, on the one hand, and the NNWS on the other hand. The effects of minimizing this composite gap are assumed to be a reduced nuclear arms competition between the NWS — particularly between the USA and the USSR — and a reduced role for nuclear weapons in the foreign policy of all NWS. These effects are viewed by advocates of the Low Posture Doctrine as valuable objectives in themselves. They are also viewed as necessary means of satisfying many of the stated military security and political prestige objectives of the NNWS, and thereby of minimizing future nuclear proliferation[24, 66].

Achievement of the military objectives of NNWS

With respect to the military security objectives of various NNWS, advocates argue that the Low Posture Doctrine will directly reduce the threat to NNWS of nuclear attack or blackmail by a hostile major NWS — and thereby achieve objective 1 — through three types of policy instrument. First, quantitative and qualitative limitations and subsequent reductions of the strategic and tactical nuclear weapon inventories of the United States and the Soviet Union will reduce both the physical threat and the probability of a nuclear attack on a NNWS. With limited nuclear inventories, each of the two major NWS must still place its primary emphasis upon maintaining an invulnerable second-strike capability in order to deter nuclear attack upon itself by the other major NWS and by minor NWS. Thus a major NWS would have a lesser and declining proportion of its strategic nuclear capability available for limited strategic nuclear attacks upon NNWS. Furthermore, if the two major NWS substantially reduce their tactical nuclear weapon capabilities, certain NNWS, particularly in Europe, would have a reduced incentive to acquire nuclear weapons for deterrence and defence.

Second, should the USA and the USSR accept certain deployment and use constraints, NNWS would be further protected from nuclear attack or blackmail by the hostile major NWS in question. Non-deployment zones would protect affected NNWS from at least early resort to nuclear weapons. Respect for existing nuclear-free zones and undertakings to respect subsequently established nuclear-free zones in substantial regions of the world would protect regional NNWS from nuclear intervention by the two major NWS in the affected regions. The provision of negative security guarantees to at least certain NNWS would to some degree protect those NNWS from actual or threatened nuclear attack. Despite their merely logistical and/or declaratory nature at the outset, such constraints upon the major NWS' deployment and use of nuclear weapons would, over time, become increasingly binding upon decision-makers as they remained in force. Given such increased self-enforcement, the amount of protection which such deployment and use constraints provide NNWS could increase over time.

Third, the Low Posture Doctrine suggests that attempting to reduce perceived threats may prove more productive than attempting to meet such threats[73a]. The USA and the USSR should not only attempt to improve their own bilateral relations in order to minimize the probability of nuclear war; they should also assure their respective close military allies that the threat of nuclear attack by the other major NWS need not be exaggerated. For example, if the USA de-emphasized the likelihood that the WTO would threaten to initiate nuclear attacks upon NATO, the USA might encourage such threat reassessment on the part of FR Germany. Each major NWS might couple such threat assessment with reaffirmation of its permanent political commitment to the NNWS in question. By so identifying certain NNWS as among its most central security interests, each major NWS might also project to the other a clearly demarcated sphere of security interest and thus deter the other from engaging in already improbable military hostilities. This dual effort might reduce the threat of nuclear attack by a hostile major NWS perceived by some NNWS.

However, it must be assumed that some NNWS would still perceive some threat of nuclear attack by a major NWS. In these residual cases, such advocates of a modified Low Posture Doctrine as Beaton, Maddox, Singer and Smart support the provision of necessarily unilateral positive security guarantees by one major NWS guarantor to NNWS allies subject to actual or threatened nuclear attack by the other[9b, 34, 60, 66]. According to these strategists, the USA should continue to serve as guarantor to its NATO NNWS allies and to Japan, and the USSR should continue to serve as guarantor to its WTO allies against nuclear attack or threatened attack by the other power. The Low Posture Doctrine should thus not be pursued to a point which would jeopardize the ability of the USA and the USSR to provide such guarantees. Each should retain a capability for extended deterrence with which to execute limited strategic responses against the other in the event of nuclear attack upon guaranteed NNWS in addition to retaining an assured destruction capability for deterring nuclear attack upon its own homeland. Such a requirement sets a floor below which strategic nuclear weapon inventories should not fall. This floor may well be above some of the traditional formulations of a minimum deterrent capability, although advocates of the modified Low Posture Doctrine contend that it is well below current US and Soviet strategic force levels[84, 85].

Furthermore, constraints upon the use of nuclear weapons should not preclude the two major NWS from providing such guarantees. No-first-use agreements should provide that nuclear attacks by one major NWS upon guaranteed NNWS constitute the prior use of nuclear weapons legitimizing a nuclear response against it by the guarantor major NWS. Similarly, negative security guarantees to NNWS should be so conditioned that, if a nuclear attack by one major NWS either occurs in the course of a conventional attack joined by NNWS allies or utilizes nuclear weapons stored on the territory of NNWS allies, such NNWS allies of the hostile major NWS are no longer protected by existing negative security guarantees from nuclear retaliation by the guarantor.

However, advocates of the Low Posture Doctrine in its extreme form, such as Falk and recently Bull, reject positive security guarantees to NNWS as both incredible and undesirable means of deterring actual or threatened nuclear attack by a hostile major NWS. They would instead rely entirely upon substantial nuclear arms limitation and reductions, non-deployment zones, nuclear-free zones and unconditional negative security guarantees to NNWS as appropriate means by which NNWS might be protected from such nuclear attack. Furthermore, to deal with the difficult case of Western Europe, Falk is prepared to countenance the development of independent West European nuclear capabilities as preferable to continuing US positive security guarantees to NATO NNWS allies as means of deterring actual or threatened nuclear attack by the WTO[40a, 73b].

Despite this disagreement about residual positive security guarantees, all advocates of the Low Posture Doctrine assume that nuclear arms limitation and reductions and various deployment and use constraints are at least necessary initial policies which the USA and the USSR must undertake in order to provide NNWS with some alternative means of achieving the nuclear component of objective 1. The provision of reasonably credible unilateral positive security guarantees would constitute an additional means of achieving the nuclear component of this objective.

Advocates admit that the Low Posture Doctrine may not contribute to the deterrence of an actual or threatened conventional attack on NNWS by a hostile major NWS. Indeed, some strategists concede the possibility that nuclear arms limitations and reductions, non-deployment zones, nuclear-free zones, and/or no-first-use agreements actually encourage such conventional attacks upon NNWS. This may be particularly true in cases, such as Western Europe, where nuclear capabilities and explicit doctrinal reliance upon the first use of nuclear weapons now contribute to a balance of otherwise asymmetric military capabilities. Thus, the Low Posture Doctrine, by failing to provide unilateral positive security guarantees, may encourage NNWS subject to conventional attack by a hostile major NWS to acquire independent nuclear weapon capabilities. Falk explicitly acknowledges this possibility with respect to West European nuclear capabilities.

While granting this possibility, all proponents of the Low Posture Doctrine argue that the threat of such conventional attacks upon NNWS by either major NWS is low and is likely to remain so. The Low Posture Doctrine explicitly proposes that the USA and the USSR undertake negotiations and agreements to limit and reduce nuclear arms and to constrain their deployment and use, as well as general efforts to improve their bilateral relations. Thus, neither of the two major NWS would be likely under the Low Posture Doctrine to initiate an actual or threatened conventional attack on a NNWS which constituted one of the central security interests of the other or to permit potentially dangerous confrontations to escalate into such attacks.

However, it must be assumed that some NNWS would still perceive some threat of conventional attack by a hostile major NWS. If such a threat were to materialize, the Low Posture Doctrine asserts that it be met entirely by non-nuclear means. Many advocates of the Low Posture Doctrine thus support the maintenance of substantial indigenous conventional forces for at least initial deterrence or defence against conventional attack by a major NWS and its allies[66]. In addition, the Low Posture Doctrine advocates various alternative policy instruments which can substitute for positive security guarantees in conventional crises between a major NWS and a NNWS. Thus, it calls for developing diplomatic capacities for the peaceful resolution of disputes; diplomatic pressure; the provision of military assistance to the threatened NNWS; and finally, the maintenance of conventional military capabilities by existing alliances for intervention in a crisis[34]. In sum, the Low Posture Doctrine asserts that despite the absence of positive security guarantees, both the low threat and the availability of alternative non-nuclear policy instruments may provide NNWS with reasonably credible alternative means of achieving the conventional component of objective 1.

More generally, it is assumed by its advocates that the high degree of cooperation between the USA and the USSR which the Low Posture Doctrine requires will, in turn, create a trend towards non-nuclear and increasingly non-military modes of resolving international conflicts. The stakes which each major NWS would have in maintaining this trend would be likely to increase over time, thereby further reducing the threat and providing NNWS with alternative means of achieving both the nuclear and conventional components of objective 1.

In the case of objective 2, just as in the case of objective 1, advocates argue that the Low Posture Doctrine can directly reduce the threat to NNWS of nuclear attack or nuclear blackmail by minor NWS in three ways. First, the Low Posture Doctrine prescribes that all minor NWS join with the major NWS in multilateral logistical and declaratory agreements establishing certain non-deployment zones, respecting existing nuclear-free zones and any subsequently established nuclear-free zones in substantial regions of the world, and providing negative security guarantees to certain NNWS. Through such deployment and use constraints, NNWS would derive protection from nuclear attack by minor NWS, which would increase as the agreements were observed over time[65d].

Second, the minor NWS should join with the major NWS in negotiations and agreements to limit and reduce their strategic and tactical nuclear weapon inventories at some point after SALT II. Such multilateral arms limitation and reduction agreements would reduce the magnitude of the nuclear threat posed by minor NWS to NNWS since much of the minor NWS's residual nuclear inventories would have to be reserved to deter nuclear attack by one or both of the major NWS[65d].

Third, the Low Posture Doctrine emphasizes the value of more accurate threat assessment. For example, by de-emphasizing the likelihood that a minor NWS such as China would threaten or initiate nuclear attacks and by avoiding such over-reactions as the 1967 US proposal to deploy an ABM system against China, the major NWS might encourage more realistic threat assessment on the part of NNWS which perceive themselves threatened by a minor NWS[34, 60, 73a].

However, it must be assumed that some NNWS would still perceive some threat of nuclear attack or nuclear blackmail by a minor NWS. To meet such perceived threats, just as in the case of objective 1, certain advocates of a modified Low Posture Doctrine — Beaton, Maddox, Singer and Smart — support the provision of positive security guarantees to allied or non-aligned NNWS subject to actual or threatened attack by a minor NWS[9b, 66]. Four conditions must be met for such guarantees to be effective. First, given the existing state of mutual deterrence between the USA and the USSR, such guarantees would be least credible if unilateral and most credible if cooperatively executed, either tacitly or explicitly, by the two countries — perhaps in conjunction with other minor NWS — against the hostile minor NWS. Second, such cooperative guarantees should be broader and more binding than the existing Security Council Resolution

255. In this view, the NWS should, through a multilateral agreement among themselves, extend such guarantees to all NNWS, or at least to all NNWS party to the NPT. Having subscribed to such an undertaking with other NWS, one NWS would not find it easy to waive its role as guarantor in a crisis for fear of alienating the hostile minor NWS. Third, such guarantees require the nuclear arms limitation and reduction by the USA and the USSR should not proceed to a point at which minor NWS enjoy substantially augmented nuclear weapon capabilities relative to the two major NWS, thereby jeopardizing effective positive security guarantees. The major NWS must each remain capable of threatening a minor NWS with an unacceptable level of damage. Thus, if any substantial arms limitation and reduction is undertaken by the two countries, it must be accompanied by comparable reductions by all minor NWS. Finally, in order for such guarantees to be credible, counterthreats by the hostile minor NWS must be deterred. Thus, intelligence and physical security measures should be such as to minimize the possibility of clandestine implacement of even a few nuclear explosive devices by a hostile minor NWS in the territory of the NWS guarantors. If these four conditions can be met, the resultant positive security guarantees would minimize future nuclear proliferation by providing NNWS with reasonably credible alternative means of achieving the nuclear component of objective 2[65a].

However, other advocates of the Low Posture Doctrine in its extreme form — Falk and recently Bull — reject positive security guarantees as a means of deterring or defending against nuclear attack or blackmail by any NWS, including minor NWS. These strategists advocate observance by minor NWS of various deployment and use constraints such as non-deployment zones, existing and subsequently established nuclear-free zones, and unconditional negative security guarantees as appropriate means by which NNWS might be protected from nuclear attack by minor NWS.

Despite this disagreement about residual positive security guarantees, all advocates of the Low Posture Doctrine assume that immediate adherence to various multilateral deployment and use constraints and eventual participation in nuclear arms limitation and reduction agreements are at least necessary policies which the minor NWS must undertake in order to provide NNWS with some alternative means of achieving the nuclear component of objective 2. The provision of reasonably credible, cooperative positive security guarantees would constitute an additional means of achieving the nuclear component of objective 2.

As in the case of objective 1, the Low Posture Doctrine rejects positive security guarantees as a means of deterring actual or threatened conventional attack on a NNWS by a local or regional minor NWS. In doing so, advocates admit that the Low Posture Doctrine may not deter an actual or threatened conventional attack on a NNWS by a minor NWS. NNWS may fear that a minor NWS would launch a conventional attack buttressed by its ability to threaten subsequent nuclear attack in order to enforce an outcome favourable to itself. Examples of such perceived conventional threats would be a conventional attack by China against India, at least up to May 1974, and a conventional attack by India against Pakistan since May 1974[36b]. Thus with respect to both objectives 1 and 2, the disinclination of NWS to use nuclear weapons first and thereby to provide positive security guarantees to NNWS in the case of an actual or threatened conventional attack by a hostile NWS may encourage NNWS subject to such attacks to acquire independent nuclear weapon capabilities.

While granting this possibility, advocates of the Low Posture Doctrine prescribe, as in the case of objective 1, a variety of non-nuclear means by which to deter or retaliate against an actual or threatened conventional attack by a minor NWS. These include initial indigenous conventional military resistance by the NNWS in question. Advocates of the Low Posture Doctrine argue in this connection that the NATO allies, by relying upon the first use of nuclear weapons as a means of deterring or defending against conventional attack by the WTO, are establishing a general system-wide precedent for NNWS in different regions to acquire nuclear weapons in order to deter, retaliate for, or defend against conventional attack by a minor NWS. Through their adoption of a no-first-use agreement, the NWS might create an alternative precedent which would steer NNWS towards conventional and away from nuclear responses to conventional attacks by minor NWS. Other non-nuclear policy instruments by which the international system in general and the NWS in particular might collectively protect NNWS from actual or threatened conventional attacks by a minor NWS include the use of international mechanisms for the peaceful resolution of disputes; diplomatic pressure; the provision of military assistance to the threatened NNWS; the interposition of international peacekeeping forces; and, finally, conventional military intervention. Thus, despite the absence of positive security guarantees, such measures may provide NNWS with reasonably credible alternative means of achieving the conventional component of objective 2.

With respect to objective 3, as in the case of objectives 1 and 2, the Low Posture Doctrine rejects positive security guarantees as being both an incredible and an undesirable means of deterring conventional attack on a NNWS by a hostile local or regional NNWS or group of NNWS. However, its advocates argue that the Low Posture Doctrine can nonetheless effectively prevent NNWS from acquiring nuclear weapons in order to deter, dominate or defend against conventionally armed adversaries and can effectively protect NNWS should such conventional attacks occur. To those ends, the Low Posture Doctrine consistently emphasizes the value of permanent multilateral deployment and use constraints which would prohibit any contemplated use of nuclear weapons except for deterring or retaliating against nuclear attack. In particular, the Low Posture Doctrine rejects the use or threatened use of nuclear weapons against NNWS. The existence of negative security guarantees to certain NNWS would put great pressure on any NNWS contemplating the acquisition of nuclear weapons in order to deter, dominate or defend against future conventional attack by hostile NNWS. First, such constraints would suggest to a NNWS that the international security system and particularly the NWS would view as highly suspect any new nuclear weapon acquisition programme explicitly intended for use in otherwise conventional contingencies against NNWS. Second, such constraints would suggest that, given the increased legitimacy of the nuclear firebreak, the threat to use nuclear weapons in contingencies which would otherwise remain conventional would be of declining credibility. Third, if coupled with the extension of cooperative positive security guarantees to NNWS, such constraints suggest that the NWS would be likely to punish any breach of the existing negative security guarantees by a new NWS should it use its nuclear weapons in an otherwise conventional contingency[27, 65d]. Thus, the existence of negative security guarantees and possible cooperative positive security guarantees might deter NNWS from acquiring nuclear weapons as a means of achieving objective 3.

In terms of protecting NNWS from conventional attack by hostile local or regional NNWS, the Low Posture Doctrine, as in the case of objectives 1 and 2, supports various non-nuclear policy instruments. As a long-term preventive measure, the international security system might work to establish not only nuclear-free zones in regions of potential hostility but also zones of peace or demilitarized zones in which conventional military capabilities would likewise be constrained. Limitation on the transfer of conventional arms would be another important policy instrument. Should conventional hostilities nevertheless threaten to occur in such instances as an Arab threat to Israel, a North Korean threat to South Korea, or a threat by a group of Black African states to South Africa, indigenous conventional military resistance would be possible[66, 81]. Major states in the international security system might also utilize international mechanisms for the peaceful resolution of disputes, diplomatic pressure, military assistance to the threatened NNWS, the interposition of international peace-keeping forces, and, finally, conventional military intervention as appropriate means of deterring or responding to such threats by hostile NNWS. Thus, under the Low Posture Doctrine, the combination of negative security guarantees, possible cooperative positive security guarantees, and the availability of non-nuclear policy instruments may combine to deter NNWS from acquiring nuclear weapons and to provide them with reasonably credible alternative means of achieving objective 3.

Fourth and finally, advocates argue that the Low Posture Doctrine can be effective in deterring either party to an incipient local or regional arms race from acquiring, deploying and using nuclear weapons in order to deter or dominate its regional adversary. The Low Posture Doctrine prescribes several policy instruments to meet objective 4. First, as a long-term means of preventing incipient local or regional nuclear arms races, advocates of the Low Posture Doctrine, particularly John Maddox, emphasize the significance of establishing nuclear-free zones in substantial regions of the world[60b, 86]. Such incipient arms races by definition occur among pairs or clusters of local or regional adversaries, many of which have not as yet signed or ratified the NPT. It has become clear in UN debates on various proposals over the past few years that before nuclear-free zones subject to International Atomic Energy Agency (IAEA) and zonal safeguards can be established in such regions as the Middle East, Africa and South Asia, major political disputes must be at least partially resolved. However, as Maddox argues, such zones would be highly effective means of defusing incipient local or regional arms races. By renouncing the acquisition, deployment and use of nuclear weapons in a given region and instituting regional safeguards, a nuclear-free zone agreement could protect potentially hostile neighbours from mutual fear of nuclear acquisition by the other. Thus it both deters a NNWS from acquiring nuclear weapons as an anticipatory reaction to their acquisition by a local or regional adversary, and deters such an adversary from acquiring nuclear weapons.

Second, as in the case of objective 3, in order to deter either party to an incipient local or regional arms race from acquiring an independent nuclear weapon capability, the Low Posture Doctrine consistently emphasizes the value of negative security guarantees to certain NNWS. Since the preemptive acquisition of nuclear weapons in such an arms race would be at least initially directed against a NNWS, negative security guarantees to NNWS would suggest to any NNWS contemplating such acquisition that the international security system, and particularly the NWS, would view it as highly suspect.

Third, if coupled with the extension of cooperative positive security guarantees to NNWS, such constraints suggest that the NWS would be likely to punish any breach of existing negative security guarantees by a new NWS were it to use its nuclear weapons against a local or regional NNWS adversary.

Finally, should such an incipient local or regional arms race occur, major states in the international system might apply a variety of non-military policy instruments in order to maintain peaceful relations between the adversary states. These might include international mechanisms for the peaceful resolution of disputes, diplomatic pressure, limitations on the transfer of conventional arms, particularly upon dual-capable delivery vehicles, the interposition of international peacekeeping forces, and the introduction of international or bilateral verification capabilities. Thus, under the Low Posture Doctrine, the combination of nuclear-free zones, negative security guarantees, possible cooperative positive security guarantees, and the availability of non-military policy instruments may combine to deter both parties to an incipient local or regional arms race and to provide them with reasonably credible alternative means of achieving objective 4.

With respect to these four military security objectives, the Low Posture Doctrine operates initially to minimize the actual and perceived threats to NNWS emanating from the nuclear weapon inventories of the existing NWS. It does so by advocating substantial limitations on and reductions of these inventories and multilateral constraints upon the permissible deployments and uses of the remaining inventories. The Low Posture Doctrine also seeks to deter NNWS from acquiring independent nuclear capabilities by stigmatizing in advance such new NWS as highly suspect and bringing pressure to bear on any new NWS to subject itself immediately to existing multilateral constraints upon permissible deployments and uses of its nuclear weapons. Furthermore, the Low Posture Doctrine offers NNWS a range of alternative non-nuclear policy instruments by which to achieve their various military security objectives. Only in residual cases involving deterrence of or retaliation against nuclear attack do advocates of the modified Low Posture Doctrine contemplate the extension of unilateral or cooperative positive security guarantees as an additional means of satisfying the nuclear components of objectives 1 or 2.

Achievement of the political objectives of NNWS

With respect to the political prestige objectives of the NNWS, the Low Posture Doctrine explicitly rejects the acquisition of independent nuclear weapon capabilities as a means of enhancing political prestige in relation to any referent group in the international political hierarchy which the NNWS deem important: whether it be the great powers, as in objective 5; existing alliances, as in objective 6; or particular local groupings, regional groupings, or transregional status cohorts, as in objective 7. It also rejects the acquisition of independent nuclear weapon capabilities by several NNWS as a means of altering the existing distribution of power and status in the international political hierarchy, as in objective 8.

Instead, the Low Posture Doctrine emphasizes the limited utility of nuclear weapons and attempts to stigmatize them as non-prestigious and politically suspect military capabilities comparable to chemical and biological weapons. By limiting and reducing nuclear weapons and by constraining their permissible deployments and uses, the Low Posture Doctrine imposes special obligations and responsibilities upon existing NWS. Indeed a Comprehensive Test Ban, no-first-use undertakings and the provision by NWS of negative security guarantees to NNWS would all subject NWS to the same behavioural norms as NNWS despite their possession of nuclear weapons[40b, 73b].

The Low Posture Doctrine assumes that a composite trend of such restrictive limits, reductions and deployment and use constraints will, over time, minimize the status differential between NWS and NNWS which is attributable to the former's possession of nuclear weapons. Thus, such attempts to delegitimize nuclear weapons have utility for NNWS in terms of gains in relative political prestige as well as in terms of military security. In addition, by minimizing the demonstration effect that nuclear weapons necessarily impart political prestige, the Low Posture Doctrine denies NNWS the convenient excuse that nuclear weapons are a necessary accoutrement of any substantial nation-state in a technological age.

The Low Posture Doctrine does not and cannot obliterate all status differentials between NWS and NNWS attributable to the possession of nuclear weapons. Nuclear weapons do have some intrinsic prestige, and as long as some states possess them and some do not, status differentials are inevitable and are likely to be perceived by the nuclear 'Have-Not' states as discriminatory[27]. However, with explicit efforts to stigmatize nuclear weapons, the Low Posture Doctrine may succeed in minimizing such status differentials[45]. Furthermore, the Low Posture Doctrine proposes some alternative means by which NNWS can enhance their political status. These effects of the Low Posture Doctrine upon the NNWS' pursuit of each of the four political prestige objectives are briefly reviewed below.

First, the Low Posture Doctrine seeks to minimize the symbolic political importance of a nuclear weapon capability as a component of great power status in the international political hierarchy and to refute the assumption that nuclear weapons are "essential to the prestige and standing of a major power"[65a]. The Doctrine thus flies in the face of post-1945 historical developments, since the USA, the USSR, the UK, France and China — all states which have been traditionally viewed as great powers by several criteria apart from nuclear weapons — have felt impelled to organize their military security explicitly around such weapons and to use them as means of exercising international influence and winning international prestige[34e]. So long as such traditional great powers continue to rely upon nuclear weapons in their conduct of foreign policy, NNWS with plausible claims to great power status are encouraged to acquire such capabilities themselves in order to achieve objective 5.

Thus, according to the Low Posture Doctrine, efforts to reduce the political prestige attached to nuclear weapons must be initiated by the existing NWS. Such efforts to deny the relevance of nuclear weapon capabilities to great power status would not be entirely selfless on the part of the existing NWS — all of which, with the possible exception of India, have traditionally ranked as great powers quite apart from their possession of nuclear weapons[34e, 66]. Should substantial nuclear proliferation occur in the future, it will to at least some degree result in relative status loss in the international political hierarchy for existing NWS relative to future NWS which, on other criteria, might be considered 'lesser' powers than the existing NWS. In any event, if restrictions upon existing nuclear weapons can make such capabilities appear non-prestigious and indeed even politically suspect, NNWS which might plausibly lay claim to great power status will have reduced incentives to acquire them as a means of achieving objective 5.

In addition to minimizing nuclear weapons as a component of great power status through limiting and constraining the nuclear capabilities of existing NWS, the Low Posture Doctrine suggests that other criteria be emphasized in identifying great powers both in the conduct of foreign policy by individual nation-states and in the organization and conduct of international institutions[26a, 34e]. For example, US foreign policy in the late 1960s and early 1970s has been criticized as excessively preoccupied with China, in part because of its nuclear capabilities, and insufficiently attentive to important policy issues of interest to Japan and India: both, in that period, important NNWS. The identity of the permanent membership of the UN Security Council with the first five NWS has often been criticized as an unfortunate coincidence which conferred undue status upon the NWS. Instead, as Leonard Beaton has suggested, international institutions such as the Security Council should recognize and incorporate important NNWS such as Brazil, FR Germany, Japan and India — at least prior to May 1974 — which can plausibly lay claim to emergent great power status by such criteria as population, wealth, historical-cultural tradition and non-nuclear forms of military power. Similarly, negotiations on the limitation, reduction and permissible uses of nuclear weapons should not be restricted to some or all of the NWS, as in SALT and in various proposals for arms limitation conferences among all NWS, but should in some way provide a role for important NNWS as well[60b, 66]. In short, the Low Posture Doctrine suggests that the NWS abandon claims to special diplomatic treatment or status as a function of possessing nuclear weapon capabilities and that there be increased attention paid to and increased participation by important NNWS in the international security system. By so concentrating upon alternative components of great power status and minimizing the significance of the nuclear weapon component, the Low Posture Doctrine will provide NNWS which seek great power status with alternative means of achieving objective 5.

Second, advocates of a modified Low Posture Doctrine support the extension of unilateral positive security guarantees to NNWS allies as a means of their achieving the nuclear component of objective 1. Such guarantees may, however, appear threatening to the political status of certain NNWS. A proud NNWS ally may see unilateral guarantees as imposing a formally inferior and possibly demeaning status as a protectorate of the dominant NWS in a military alliance. A proud non-aligned NNWS may wish to assert its non-aligned status and political independence from existing military alliances.

To the degree that the Low Posture Doctrine, by limiting and constraining the nuclear capabilities of existing NWS, can reduce the symbolic political importance of nuclear weapons within the international security system, it will also reduce the incentives of NNWS — both allied and non-aligned — to acquire nuclear weapons in order to gain political status *vis-à-vis* existing military alliances and thus to achieve objective 6. It would no longer seem necessary for a NNWS to acquire nuclear weapons in order to assert an independent status either within or explicitly outside a military alliance, since such weapons would no longer automatically confer enhanced political prestige.

Furthermore, if any cooperative positive security guarantees provide for enhanced participation by the NNWS which are the recipients of such guarantees through the processes which Leonard Beaton has termed cognizance and commitment (see page 50), these NNWS will have some access to whatever residual political prestige nuclear weapons impart[42d]. If NWS guarantors were to offer such a participatory role, it could serve as an alternative means by which NNWS could achieve objective 6 as well as the nuclear components of objectives 1 and 2.

Finally, if no positive security guarantees against actual or threatened nuclear attack exist — as advocated by Falk and recently by Bull — there would be no formal or tacit differentiation within or outside of military alliances between NWS guarantors and the NNWS which might receive such guarantees. Thus, no invidious status distinctions would arise to create in NNWS a perceived need for rectification, and objective 6 would become moot.

Third, to the degree that the Low Posture Doctrine, by limiting and constraining the nuclear capabilities of existing NWS, can reduce the symbolic political importance of nuclear weapons within the international security system, it will also reduce the incentives of NNWS to acquire nuclear weapons in order to gain second-order power status and thus achieve objective 7. It would no longer seem necessary for a NNWS to acquire nuclear weapons in order to enhance its political prestige in a particular local grouping, regional grouping, or transregional status cohort, since such weapons would no longer automatically confer enhanced political prestige. Furthermore, it is consistent with the Low Posture Doctrine to encourage the current emergence of several NNWS as claimants to second-order power status in the international political hierarchy. Increased receptivity by the existing NWS to NNWS' playing important roles in various nuclear, non-nuclear and non-military sectors of international politics would provide alternative means by which such NNWS could achieve objective 7.

In the previous three cases, of objectives 5, 6 and 7, the Low Posture Doctrine attempts both to diminish the incentives and to provide alternative policy instruments for NNWS which may be contemplating the acquisition of nuclear weapons as a means of achieving enhanced political prestige within the existing international political hierarchy or some subset thereof. Objective 8 challenges the hierarchy itself. If the Low Posture Doctrine can de-emphasize the military and political utility of nuclear weapons and, indeed, stigmatize them as non-prestigious and politically suspect, the widespread possession of an independent nuclear weapon capability ceases to be an obvious political symbol around which several NNWS might challenge the NWS and the other industrialized powers which currently dominate the international political hierarchy. Furthermore, such stigmatizing can prevent the widespread proliferation of nuclear weapons which might ensue if, after several NNWS had previously acquired them, many other NNWS were to feel pressed to follow suit.

Perhaps most significant in this regard is that the Low Posture Doctrine creates a quite comprehensive arms limitation and security régime which amplifies the NPT by imposing counterdiscriminatory obligations and risks upon the NWS. It thus rectifies what the NNWS have perceived as a discriminatory allocation of rights and obligations inherent in the NPT and substantially diminishes the tacit NWS hegemony legitimized by the NPT. To quote Lincoln Bloomfield on this point:

> In an era dominated by demands for identity, respect, equity, and participation, it seems reasonable to ask whether, with the best will in the world, the present NPT system of discrimination, denial and second-class citizenship will in fact achieve its aim of preventing the further spread of nuclear weapons. For if my reasoning is correct, it is considerations of prestige and non-discrimination that in an age of rampant nationalism stand as the chief obstacles to universal agreement on proliferation.
>
> The logic of the situation requires that an expanded nonproliferation strategy focus on tangible ways to give the outsiders a far more genuine sense of participation in the system. It is obviously not enough to say to them, with the late President Kennedy, "Life is unfair". Deep-rooted feelings of political alienation and resentment can be overcome only by greater true equality. There must be shared opportunities to gain prestige through participation in decision-making, which in turn requires that responsibilities be much more broadly allocated than under the present two-class system. The operative hypothesis is that a seat at the top table of nuclear institutional diplomacy is the price the 'monopolists' must pay to others who agree to forgo a seat at the top table of nuclear weaponry[27, 60c, 61a].

Moreover, it is particularly necessary that the NWS pursue counterdiscriminatory arms limitation and security policies, since current nuclear supply policies and existing safeguards systems to some degree inherently discriminate against the economic interests and political independence of the NNWS.

By encouraging participation of NNWS in various non-nuclear and non-military as well as nuclear sectors of international politics as alternative means of achieving objectives 5, 6 and 7, the Low Posture Doctrine has already gone some distance towards meeting the aspirations of NNWS currently dissatisfied with the existing international political hierarchy. If the NWS, and particularly the USA and the USSR, could undertake counterdiscriminatory obligations, share power and encourage participation in the international security system, power and the right to participate would not have to be wrested from them by several NNWS and objective 8 might to some degree become moot.

In sum, the Low Posture Doctrine seeks to stigmatize nuclear weapons and thereby minimize their utility as a symbol of enhanced political prestige. The Low Posture Doctrine assumes that dominance of the international security system by NWS is to some degree inevitable and residual status differentials between NWS and NNWS are inescapable. However, the Low Posture Doctrine does argue that minimizing these status differentials will reduce the incentives for NNWS to acquire nuclear weapons in order to gain enhanced political prestige. It argues further that more widely shared power and participation in the international system may create common stakes in a non-nuclear world order, which in turn may substantially satisfy the political prestige objectives of many NNWS. The NWS would in all likelihood continue to dominate the international political hierarchy, but theirs would be a far more benign, and thus more acceptable, central role. To quote Hedley Bull on this concluding point:

> There is a certain justice in the note of grievance which is sometimes struck by countries which see themselves as the nuclear Have-Nots or proletarians. It is true that in an international order in which the many do not have nuclear weapons, the few that retain them will enjoy privileges, however effectively they are able to disguise them. But the alternative to an international order in which certain states have a larger stake than others is probably no international order at all. The problem is not to find an international order in which no one state or group of states has a special interest, but rather to ensure that those who do have special interests recognize the special responsibilities that go with them, and conduct themselves in such a way as to engage general support for the system whose custodians and guarantors they are. It is in this latter sense that the doctrine of Low Posture is most defensible[65a].

III. Conclusion

A summary assessment of the High Posture and Low Posture Doctrines suggests that the adoption of a modified Low Posture Doctrine by the NWS would best satisfy most of the various policy objectives of the NNWS, thus amplifying the NPT and minimizing future nuclear proliferation.

An assessment of the *High Posture Doctrine* as an anti-proliferation strategy must give it a mixed but potentially fairly low rating. The primary policy instrument which the High Posture Doctrine offers in both the extreme and the Frye versions is the extension by the major NWS of unilateral or cooperative positive security guarantees to various NNWS. This policy instrument is, however, an increasingly ineffectual and possibly counterproductive anti-proliferation strategy, as positive security gurantees suffer an irreversible decline in credibility and acceptability. Given the state of mutual deterrence existing between the USA and the USSR, unilateral positive security guarantees are a 'depreciating asset' for NNWS[27, 36a]. At least in the USA since the end of the war in Indo-China, such international commitments may be suffering a long-term erosion of domestic political support. Indeed, given their declining credibility, the emphasis placed on them in the High Posture Doctrine as the primary policy instrument by which proliferation is to be contained may become counterproductive, if NNWS are reminded too frequently of their relative inefficacy. Moreover, the value of cooperative positive security guarantees in the relatively few contingencies in which US-Soviet cooperation would be credible under the High Posture Doctrine is depreciating as opposition to such great power domination renders them politically less acceptable. Thus, inevitably, "guarantor's stock is going down,[and] fledgling nuclear stock up on the international power exchange"[36a, 67c].

The rate of this depreciation in credibility will vary. With respect to deterring nuclear or conventional attack or nuclear blackmail by the other major NWS against NNWS allies, the centrality of the security interests at stake and the availability of limited nuclear response options combine to provide unilateral positive security guarantees of substantial, although declining, credibility. The High Posture Doctrine also raises the costs for NNWS of acquiring nuclear weapons to be directed against a potentially hostile major NWS. It thus serves as a reasonably effective alternative means of satisfying and deterring NNWS which are contemplating the acquisition of an independent nuclear weapon capability in order to achieve objective 1. This is the primary anti-proliferation objective of the High Posture Doctrine: indeed, the cohesion of the US alliances with NATO and Japan and, necessarily, the cohesion of the WTO, are thought by some of its proponents to be of higher priority than the overall minimization of future nuclear proliferation. Thus it is objective 1 which the High Posture Doctrine achieves most effectively.

With respect to deterring nuclear or conventional attacks by local or regional adversaries — either minor NWS or NNWS — the High Posture Doctrine provides NNWS with fairly credible positive security guarantees or commitments to intervene with conventional forces only if the security interest of one major NWS is sufficiently unimportant to the other so that it actually tolerates intervention by the interested power, or if important security interests of both major NWS converge, enabling them to intervene jointly in a local or regional crisis. In these contingencies, the High Posture Doctrine can provide NNWS with reasonably effective alternative means of achieving objectives 2, 3 and 4. However, as noted above, there is rising opposition to such great power dominance in the international security system. Thus, the hegemonic character of such cooperative guarantees — whether direct or indirect — and conventional commitments may render them politically unacceptable over time. Furthermore, the High Posture Doctrine acknowledges that positive security guarantees and commitments to intervene with conventional forces may not be credible in a local or regional crisis if the USA and the USSR are either disinterested or at odds. To the degree that the US-Soviet strategic balance is maintained at a high level in order to ensure credible unilateral positive security guarantees to their respective major allies, political tension between the two powers is likely to continue under the High Posture Doctrine. Thus, the likelihood of their joining in cooperative positive security guarantees or conventional commitments to other NNWS may remain low. In such contingencies, the High Posture Doctrine will fail to provide NNWS with an alternative means of achieving objectives 2, 3 and 4.

Furthermore, with respect to the political prestige objectives of NNWS, the High Posture Doctrine is counterproductive as an anti-proliferation strategy in that it explicitly encourages NNWS to acquire nuclear weapons in order to achieve objectives 5, 6, 7 and 8. This result is inescapable since the High Posture Doctrine contains a trade-off: the strategic superiority and extended deterrent posture which it prescribes in order to fulfil the military security objectives of certain NNWS focuses attention upon nuclear weapons and encourages other NNWS to acquire independent nuclear weapon capabilities in order to enhance their political prestige.

Thus, while providing major NNWS allies with a reasonably credible alternative means of achieving objective 1, the High Posture Doctrine leads to a fairly high probability of future nuclear proliferation by a significant number of NNWS pursuing the other seven objectives. Specifically, NNWS for which the credibility of the major NWS' positive security guarantees or conventional commitments in local or regional crises depreciates too much over time as well as NNWS pursuing enhanced political prestige may choose to exercise their nuclear weapon option. Indeed, eventually even the credibility of unilateral positive security guarantees extended to major NNWS allies in the event of nuclear or conventional attack or nuclear blackmail by a major NWS may depreciate as well.

The High Posture Doctrine tolerates this prospect of substantially increased future nuclear proliferation by assuming that the risks of nuclear proliferation will fall on the NNWS involved in local or regional conflicts. Conversely, it assumes that the costs to the USA and the USSR will be acceptable so long as they maintain their vantage point of strategic superiority over all minor NWS. To quote James Schlesinger on this point:

> In the absence of major investments or extraordinary outside assistance the only option open to most nuclear aspirants is the aerial delivery of rather crude nuclear weapons. Though such capabilities can, of course, dramatically transform a regional balance of power (provided that the superpowers remain aloof), the superpowers themselves will remain more or less immune to nuclear threats emanating from countries other than the principal opponent[36d, 65a, 65b].

Assuming that the type of nuclear capabilities acquired by NNWS involved in local or regional conflicts or arms races cannot directly threaten the USA and the USSR, the High Posture Doctrine concludes that such increased nuclear proliferation is preferable to the major NWS substantially reducing their strategic capabilities and thus putting at risk either their own security *vis-à-vis* the other or the security of their closest and most central military allies. Consequently, the High Posture Doctrine further concludes that the USA and the USSR should not undertake substantial changes in their own arms acquisition, arms limitation or security policies in order to minimize the proliferation of nuclear weapons[36d, 65b].

A summary assessment of the *Low Posture Doctrine* as an anti-proliferation strategy must give it a mixed but potentially fairly high rating, particularly in the case of the modified Low Posture Doctrine. In contrast with the limited scope of the High Posture Doctrine, the Low Posture Doctrine propounds a comprehensive range of major, mutually compatible policy instruments as means of satisfying the military security and political prestige objectives of various NNWS. Chief among these are the limitation and reduction of existing nuclear weapon inventories, particularly those of the two major NWS; constraints upon the permissible deployments and uses of the remaining nuclear weapon inventories; primary reliance upon various non-nuclear and non-military policy instruments for resolving conflicts and attributing status in the international security system; and, finally, in the modified version of the Low Posture Doctrine, ultimate reliance in certain hard residual cases upon the provision of unilateral or cooperative positive security guarantees to deter or retaliate against nuclear attack on certain NNWS. Advocates of the Low Posture Doctrine thus seek to buttress the existing NPT with a combination of collateral arms limitation and security measures, each of which, by providing alternative means of achieving various military security and/or political prestige objectives, may induce one or more NNWS to forgo the acquisition of nuclear weapons. Thus it is the cumulative impact of the multiple policy instruments advocated in the Low Posture Doctrine that may minimize future nuclear proliferation.

The combination of arms limitation and security measures included in the Low Posture Doctrine varies in the degree to which it effectively achieves the different military security and political prestige objectives of the NNWS. With respect to deterring nuclear attacks upon NNWS by either a major or minor NWS, the Low Posture Doctrine initially attempts to limit the nuclear capabilities of the NWS and to constrain their rights to deploy and initiate the use of nuclear weapons. These restrictions, if adhered to, would reduce to nil the threat of nuclear attack or blackmail against NNWS.

In the extreme version of the Low Posture Doctrine, these arms limitation and use constraints constitute the sole means by which NNWS can achieve the nuclear components of objectives 1 and 2. In this instance, a NNWS would have no recourse should an actual or threatened nuclear attack materialize. Perceiving a realistic threat of nuclear attack from a hostile NWS, in spite of the use constraints, a NNWS might thus be tempted to acquire an independent nuclear weapon capability in order to achieve the nuclear component of objectives 1 or 2.

In the modified version of the Low Posture Doctrine, the NWS extend unilateral or cooperative positive security guarantees to certain NNWS subject to actual or threatened nuclear attack by either a hostile major NWS or a minor NWS when all other policy instruments appear to have failed. Given the state of mutual deterrence and détente which would exist between the USA and the USSR under a Low Posture Doctrine, unilateral positive security guarantees to NNWS allies subject to nuclear attack by the other can never be completely credible. Indeed, as noted above in connection with the assessment of the High Posture Doctrine, their value is depreciating. However, this element of the modified Low Posture Doctrine — offering unilateral positive security guarantees to major NNWS allies as a last resort and thereby expressing a willingness to use nuclear weapons on their behalf in certain nuclear contingencies — operates in marked contrast to the broad direction of the doctrine which otherwise emphasizes explicit constraints upon the permissible uses of nuclear weapons. Thus, such specific guarantees emphasize the exceptional character of the security interests at stake and, in this way, provide such NNWS with a reasonably credible additional means of achieving the nuclear component of objective 1.

Furthermore, the extension of cooperative positive security guarantees to certain NNWS as a last resort in order to deter or retaliate against nuclear attack by a minor NWS would be more binding upon the NWS under the modified Low Posture Doctrine than Security Council Resolution 255 currently is. Given the state of détente which would exist between the USA and the USSR under the Low Posture Doctrine, such cooperative guarantees are more credible than under the High Posture Doctrine. Moreover, since such guarantees are, according to the modified Low Posture Doctrine, merely one of several explicit counterdiscriminatory obligations to be undertaken by the NWS on behalf of the NNWS, they also appear less hegemonic and more politically acceptable to NNWS than does the explicit major NWS hegemony advocated by the High Posture Doctrine. Thus, these guarantees may provide NNWS with a reasonably credible and acceptable additional means of achieving the nuclear component of objective 2.

With respect to deterring or defending against conventional attacks by either a hostile major or a minor NWS, both variants of the Low Posture Doctrine are of mixed effectiveness. Since all NWS should be party to a no-first-use agreement, positive security guarantees cannot be extended to NNWS so threatened. The Low Posture Doctrine does create expectations that any new NWS would immediately become subject to existing no-first-use undertakings. Such a use constraint is designed to deter NNWS from acquiring nuclear weapons in order to achieve the conventional components of objectives 1 and 2. The Low Posture Doctrine also advocates a variety of non-nuclear and non-military policy instruments, ranging from improved institutions for the peaceful resolution of disputes to conventional intervention by military allies, in order to provide NNWS with alternative means of achieving the conventional components of objectives 1 and 2. Only if a NNWS perceived a realistic and irreversible conventional threat from a hostile NWS and found the alternative non-nuclear instruments insufficient, might it have no other recourse than to acquire an independent nuclear weapon capability, despite the use constraints, as a means of achieving the conventional components of objective 1 or 2. This might be likely if such a NNWS — as in the case of FR Germany confronting the USSR or Pakistan confronting India — had a substantially smaller population and/or GNP than its NWS opponent and thus no reasonable expectation of being able to mount an effective conventional defence.

In the case of objective 3, both variants of the Low Posture Doctrine create expectations that any new NWS would immediately become subject to existing deployment and use constraints with respect to certain NNWS. Such constraints are designed to deter NNWS from acquiring nuclear weapons in order to achieve objective 3. The Low Posture Doctrine also advocates various non-nuclear policy instruments as alternative means of achieving objective 3. Only if a NNWS perceived a realistic and irreversible threat from local or regional NNWS adversaries and found the alternative non-nuclear instruments insufficient might it have no other recourse than to acquire an independent nuclear weapon capability, despite use constraints, as a means of achieving objective 3. This might be likely, however, if such a NNWS, as in the case of Israel confronting several Arab states, had a substantially smaller population and/or GNP than its local or regional NNWS opponents and thus no reasonable expectation of being able, over time, to mount an effective conventional defence.

With respect to objective 4, both variants of the Low Posture Doctrine advocate the establishment of nuclear-free zones in substantial regions of the world both as a long-term means of deterring NNWS from entering into a local or regional arms race and as a means of protecting them from such actions by hostile neighbouring states. As in the case of objective 3, it also creates expectations that any new NWS would immediatley become subject to other existing deployment and use constraints with respect both to certain NNWS and to other NWS. These constraints are designed to deter NNWS from acquiring nuclear weapons in order to achieve objective 4. Finally, the Low Posture Doctrine advocates various non-military policy instruments as means of maintaining peaceful relations among adversary states should such an incipient arms race actually materialize. Only if a NNWS perceived a realistic threat of prospective acquisition of nuclear weapons by a local or regional NNWS adversary and found the alternative non-military instruments insufficient might it have no other recourse than to acquire an independent nuclear weapon capability, despite use constraints, as a means of achieving objective 4.

Finally, with respect to the political prestige objectives of various NNWS, the Low Posture Doctrine in both variants minimizes future nuclear proliferation in two ways. First, it explicitly attempts to stigmatize nuclear weapons, thereby minimizing their utility for enhancing political prestige. Second, it attempts to provide NNWS with alternative means of achieving objectives 5 to 8.

Thus, the modified Low Posture Doctrine in particular can lead to a fairly low incidence of future nuclear proliferation. Under this variant, it is only when various arms limitation and use constraints fail to reduce the threats facing certain NNWS and non-nuclear policy instruments and residual positive security guarantees appear insufficient to protect these NNWS that a NNWS will have no recourse but to acquire an independent nuclear weapon capability in order to achieve its military security objectives. Furthermore the modified Low Posture Doctrine — like the extreme version — minimizes the number of additional NNWS which will seek to acquire nuclear weapons in order to gain enhanced political prestige.

As depicted in table 2.2, the modified Low Posture Doctrine dominates both the High Posture Doctrine and the extreme Low Posture Doctrine in the degree to which it satisfies the various policy objectives of the NNWS. It fulfils more effectively than does the High Posture Doctrine

Table 2.2. Effectiveness of alternative postures to satisfy various objectives of NNWS and thus minimize proliferation

Objective	High Posture Doctrine	Modified Low Posture Doctrine	Extreme Low Posture Doctrine
1[a]			
Against nuclear attack	Effective	Effective	Ineffective
Against conventional attack	Effective	Mixed	Mixed
2			
Against nuclear attack	Mixed	Effective	Ineffective
Against conventional attack	Mixed	Mixed	Mixed
3	Mixed	Effective	Effective
4[a]	Mixed	Effective	Effective
5	Ineffective	Effective	Effective
6	Ineffective	Effective	Effective
7[a]	Ineffective	Effective	Effective
8[a]	Ineffective	Effective	Effective

[a] Most common objectives of NNWS.

the regional military security objectives and the political prestige objectives of the NNWS. And it deters more effectively than does the extreme Low Posture Doctrine nuclear attacks on NNWS by existing NWS. It is thereby the comprehensive anti-proliferation régime which most effectively amplifies the NPT and promises to minimize future nuclear proliferation.

As argued throughout, the objective of a comprehensive anti-proliferation régime can only be to minimize future nuclear proliferation, not to stop it. There are likely to be certain hard, residual cases in which the modified Low Posture Doctrine will also be ineffective in preventing a NNWS from acquiring nuclear weapons either by deterring it or by providing it with alternative means by which to achieve its various military security and/or political prestige objectives.

However, the modified Low Posture Doctrine is very much worth pursuing for three important reasons. First, it may well be a necessary if not sufficient condition of minimizing future nuclear proliferation. Under the High Posture Doctrine, as argued above, substantial nuclear proliferation is likely to occur. Unless the arms limitation and security policies incorporated in the modified Low Posture Doctrine are adopted, there will be little hope of minimizing future nuclear proliferation. In short, failure to pursue a modified Low Posture Doctrine means acceptance of a high probability of substantial nuclear proliferation in the future.

Second, the modified Low Posture Doctrine would substantially raise both the domestic and international costs for NNWS of exercising their nuclear weapon option. By imposing constraints upon the existing NWS in the development, production, deployment and use of nuclear weapons, the modified Low Posture Doctrine deprives NNWS of easy and currently available excuses to acquire nuclear weapons of their own. It also places them in some military and political jeopardy if they do proceed to acquire nuclear weapons.

Finally, the modified Low Posture Doctrine would create a comprehensive arms limitation and security régime which would substantially constrain the role of nuclear weapons in the international security system. Should future nuclear proliferation materialize even after the modified Low Posture Doctrine is in place, its constraints would insulate the behaviour of any new NWS which did appear. Thus it would make a proliferated world safer than would have been the case under the High Posture Doctrine.

The modified Low Posture Doctrine would, as noted above, introduce a comprehensive range of new arms limitation and security policies. What is required primarily of the NWS in accepting these policies is the undertaking of substantial new obligations in the international security system necessitating hard choices, the adoption of new norms of international behaviour, and the acceptance of reduced freedom of action. As chapter 3 describes, many NNWS have been urging the NWS to undertake such obligations since negotiations began on the NPT in 1965.

3. Negotiations on the NPT, 1965-68

This chapter describes the argument which emerged between the NWS and the NNWS during negotiations on the NPT from 1965 to 1968 concerning the nature of the relationship between the arms limitation and security policies of the major states — particularly the NWS — and the acquisition of independent nuclear weapon capabilities by additional states. Chapter 4 describes the reconsideration of this putative relationship at the Review Conference of the NPT in 1975. Much of the argument on this issue between the NWS and the NNWS during and since the negotiations on the NPT parallels the argument between advocates of the High Posture Doctrine and the Low Posture Doctrine, respectively. It is important to focus upon these negotiations because the NPT has been the central fixture of the existing anti-proliferation régime, and its negotiation and review have been the principal occasions for questioning the appropriate design of an effective anti-proliferation régime. Thus, the theoretical relationships posited by strategists have been actively explored in these negotiating contexts.

The non-proliferation of nuclear weapons has been discussed as a component of various arms limitation and disarmament proposals since the onset of the nuclear age. During the 1940s and early 1950s, disarmament negotiations did not distinguish the non-proliferation of nuclear weapons from the elimination of nuclear weapons. The first draft resolution addressing the non-proliferation of nuclear weapons as a separate measure was introduced by Ireland at the UN in 1958[87]. Thereafter, the non-proliferation of nuclear weapons was an explicit subject of discussion at successive UN General Assemblies.

In January 1964, after the Partial Test Ban Treaty had been achieved and comprehensive negotiations on a General and Complete Disarmament (GCD) treaty had become fruitless, the USA and the USSR each proposed an agenda of collateral arms limitation and disarmament measures for subsequent negotiation[88, 89]. The four collateral measures which appeared on both agenda were (*a*) reducing the danger of surprise attack; (*b*) the freezing, reducing and/or eliminating of strategic delivery vehicles; (*c*) a comprehensive test ban; and (*d*) the non-proliferation of nuclear weapons, defined as both their non-dissemination by NWS and their non-acquisition by NNWS.

Of these four issues on which there was some shared major NWS willingness to negotiate, three were postponed or became moot. Since 1964, the danger of surprise attack has, due to advances in surveillance technology, become increasingly susceptible to unilateral, national means of verification. The control and reduction of strategic vehicles ultimately became the subject of bilateral US-Soviet negotiations at the Strategic Arms Limitation Talks. The Comprehensive Test Ban (CTB) has, since 1964, been the subject of intermittent but unsuccessful discussion at the ENDC/CCD and was transmuted by the USA and the USSR into the Threshold Test Ban Treaty of 1974.

From this shared agenda, a consensus emerged in 1964-65 between the USA and the USSR that the one remaining measure, the non-proliferation of nuclear weapons, was to be the 'next' collateral arms limitation and disarmament measure on the multilateral negotiating agenda[35c, 80b, 90]. In June 1965, the UN Disarmament Commission adopted an Omnibus Resolution DC/225 sponsored by Sweden and 28 other largely non-aligned states recommending, with specific reference to the non-proliferation of nuclear weapons in paragraph 2(c), that the ENDC "accord special priority to the consideration of the question of a treaty or convention to prevent the proliferation of nuclear weapons". Sustained negotiations upon non-proliferation thus began in the ENDC in the spring of 1965 and continued in the ENDC and the UN until the Treaty on the Non-Proliferation of Nuclear Weapons was opened for signature on 1 July 1968.

During these three years, two major questions split the states negotiating an NPT. The first concerned the indirect dissemination to NNWS of access to or control of nuclear weapons through military alliances. In sum, the question turned on whether FR Germany was to gain access to or control of nuclear weapons through NATO. It generated substantial conflict between the USA and the USSR, together with their respective allies, and largely dominated negotiations on the NPT from 1965 until early 1967. The initial disagreement concerned the US proposal for a Multilateral Nuclear Force in NATO. The USA dropped this proposal in late 1966 after the West German Social Democratic Party, which was opposed to West German acquisition of nuclear weapons, entered the 'Grand Coalition' Government. Disagreement then shifted to the hypothetical question of whether a future federated Western Europe, including FR Germany, might become a NWS through inheriting nuclear status from the national arsenal of a constituent nation-state. In 1967 the USA and the USSR 'resolved' this question through different interpretations of whether or not such a federation might so inherit nuclear status. The USSR held that a Western European Community would be a NNWS since it would be established after 1 January 1967. The Western powers held that such a Community would constitute a 'European option' and could thus become a NWS by inheriting nuclear status from France and/or the UK. Having agreed to disagree on this hypothetical issue, the USA and the USSR then agreed on language concerning the dissemination and acquisition of nuclear weapons which was incorporated in the identical draft treaties tabled on 24 August 1967.

The second major question is the focus of concern here: namely, the degree to which agreement on an NPT could be facilitated, the NPT amplified, and the anti-proliferation régime made more effective by adopting a programme of related arms limitation and disarmament measures and security guarantees[91]. This question cross-cut alignments on the first by dividing the participating NWS — the USA, the UK and the USSR, and some but not all of their respective close military allies — from many NNWS which were primarily, but not exclusively, non-aligned states. In sum, the NWS were opposed to tying related arms limitation and security measures to an NPT while the NNWS argued that an NPT was inextricably related to such measures. In November 1965, the General Assembly passed Resolution 2028 (XX), sponsored originally by the eight non-aligned NNWS in the ENDC, which summarized the main principles which the NNWS argued should guide subsequent negotiations on non-proliferation:

(*a*) The treaty should be void of any loop-holes which might permit nuclear or non-nuclear Powers to proliferate, directly or indirectly, nuclear weapons in any form;
(*b*) The treaty should embody an acceptable balance of mutual responsibilities and obligations of the nuclear and non-nuclear Powers;
(*c*) The treaty should be a step towards the achievement of general and complete disarmament and, more particularly, nuclear disarmament;
(*d*) There should be acceptable and workable provisions to ensure the effectiveness of the treaty; and
(*e*) Nothing in the treaty should adversely affect the right of any group of States to conclude regional treaties in order to ensure the total absence of nuclear weapons in their respective territories.

Thus, the major question of concern here — the nature of the relationship between the arms limitation and security policies of the NWS and the future proliferation of nuclear weapons to additional states — emerges clearly at the outset of the sustained negotiations on an NPT.

It must be granted that some NNWS, such as Brazil, India and Israel, may well have defended idiosyncratic intentions to retain a nuclear weapon option by joining in the general objections raised by other NNWS to a separate, collateral NPT. Such states are thus open to the charge of insincere, self-serving or irrelevant argumentation[49a]. Some such special cases were discussed in chapter 2. The existence of such instances does not, however, detract from the generality, plausibility and symbolic importance of the arguments made by the NNWS as a class. Many NNWS — large and small, representing all regions of the world — have argued consistently during and since the negotiations on the NPT that the NWS would have to undertake certain related arms limitation and security measures in conjunction with an NPT in order to achieve a viable treaty and to minimize the future proliferation of nuclear weapons to additional states. The NWS have consistently taken issue with this position. Thus, the conflict between these two classes of states concerning the necessity of tying related arms limitation and security measures to an NPT warrants review and analysis, regardless of the motivations of some of the particular NNWS involved.

The question of which arms limitation and security measures should be related to an NPT produced the first active collaboration between the USA and the USSR in post-war arms limitation and disarmament negotiations. By opposing the declared interests of many NNWS, the USA and the USSR and their close military allies thus instigated the first truly multilateral rather than basically bilateral negotiation on an arms limitation and disarmament measure in the post-war period. This argument dividing the NWS from many NNWS broke down into three specific issues relating to the broader arms limitation and security régime of interest to the NNWS. First, what measures of nuclear arms limitation and disarmament should be linked to an NPT and in what manner? Second, what types of security guarantee should be linked to an NPT and in what manner? And third, what procedural provisions for review, duration and withdrawal should be incorporated in an NPT lest its substantive provisions not be observed or prove inadequate over time?

These issues evolved through five stages of negotiation on the NPT between 1965 and 1968. In each stage, the USA and the USSR, on the one hand, submitted draft treaties. The NNWS, on the other hand, submitted criticisms and counterproposals. In the first stage, beginning in August 1965, the USA and the USSR each submitted separate and quite different draft treaties. The NNWS viewed each of these as inadequate and fully elaborated their contrary positions. In the four successive stages of negotiations on the NPT, beginning respectively on 24 August 1967, 18 January 1968, 11 March 1968 and 31 May 1968, the USA and the USSR submitted identical and ultimately joint draft treaties, each marginally amended to meet the continuing criticisms of the NNWS. The first four stages of the negotiations of an NPT took place primarily at the ENDC. The fifth stage of the negotiations concluded an extensive and rather bitter debate at a General Assembly session in April-June 1968 which had been specifically reconvened in order to consider the issue of non-proliferation[92].

In reviewing the argument between the NWS and the NNWS, this chapter collapses the five stages of the negotiations on the NPT together into three separate sections dealing with each of the major questions at issue between the NWS and the NNWS: (*a*) the arms limitation and disarmament measures to be linked to an NPT (*b*) the security guarantees to be linked to an NPT; and (*c*) the provisions for review, duration and withdrawal. The final section assesses the overall balance, mutual responsibilities and obligations in the NPT.

I. Arms limitation and disarmament measures

On 17 August and 24 September 1965, the USA and the USSR respectively submitted separate and different draft treaties on the non-proliferation of nuclear weapons[93, 94]. They differed, as noted above, largely in their respective treatment of what constituted dissemination and acquisition of nuclear weapons with regard to NNWS members of military alliances and in the design of an international control system. Even at this preliminary stage in the negotiations, however, the two drafts were comparable in their treatment of the three issues relating to the broader arms limitation and security régime of interest to the NNWS. The first and perhaps most central issue dividing the NWS from the NNWS concerned what measures of arms limitation and disarmament should be linked to an NPT and in what manner.

The 1965 draft treaties

Neither the US nor the Soviet draft incorporated in the body of the treaty any arms limitation and disarmament measure other than the non-dissemination of nuclear weapons by NWS and the non-acquisition of nuclear weapons by NNWS. The only references to related arms limitation and disarmament measures were in very general preambular language. The US draft referred to the desirability of "refrain[ing] from taking steps which will extend and intensify the arms race, . . . achiev[ing] effective agreements to halt the nuclear arms race, and to reduce armaments including particularly nuclear arsenals, . . . and achieving agreement on general and complete disarmament under effective international control". The Soviet draft treaty referred to the desirability of "the earliest possible attainment of agreement on the complete prohibition and elimination of all types of nuclear weapons within the framework of general and complete disarmament under strict international control".

This failure to incorporate in an NPT any related measures of arms limitation and disarmament rested upon five major arguments elaborated by the USA, the UK and the USSR during the two years of negotiations following submission of these drafts. All three argued, first, that progress in arms limitation and disarmament must be cumulative, and that failure to reach agreement on an NPT would interrupt the improvement in East-West relations generated in 1963 by the Partial Test Ban Treaty, the 'Hot Line' Agreement and the General Assembly Resolution outlawing nuclear weapons in space[95-97]. According to this argument, any single collateral arms limitation measure such as an NPT which was acceptable to the USA and the USSR — no matter how limited — was worth pursuing in order further to stabilize international security[98].

Second, the three NWS argued that the proliferation of nuclear weapons was intrinsically dangerous for both NWS and NNWS, since it would destabilize international politics. However, in an argument used in support of the High Posture Doctrine, the NWS asserted that the proliferation of nuclear weapons would be particularly dangerous for the NNWS since they would be differentially jeopardized if hostile neighbours or regional powers were to exercise nuclear weapon options. Therefore, it was in the special interest of the NNWS to reach agreement on an NPT whether or not it was linked to any other arms limitation and disarmament measures. To quote Ambassador Chalfont of the UK on this point:

> I would ask the non-aligned delegations to ponder on this point in case it turned out to be impossible to get agreement among the nuclear Powers to some measures of reduction.... I should like to ask the non-nuclear Powers most seriously whether, if this position were reached — a treaty within our grasp, but the choice of collateral measures still in dispute — it would not still be in the interest of every non-nuclear State to call a halt to the spread of nuclear weapons even if the nuclear weapon Powers themselves had not actually begun to disarm[95, 99-102].

Third, while not denying that the non-proliferation of nuclear weapons was related to other measures of arms control and disarmament, the NWS argued from the legislative history of the issue that it was clearly legitimate to negotiate an NPT as a separate collateral agreement which excluded related measures. To quote Ambassador Foster of the USA on this point:

> I should like to turn now to another type of hurdle which seems increasingly to get in our way. This hurdle is the demand that a non-proliferation treaty contain obligations on the nuclear weapons States to cease all nuclear weapon tests, to halt production of fissionable material for weapons, to stop making nuclear delivery vehicles, to put a ceiling on numbers of nuclear weapons or vehicles, or even to begin nuclear disarmament.
>
> The underlying concept of non-proliferation has been well understood for many years. Non-proliferation, like the cessation of nuclear weapon tests or the cessation of the production of fissionable material for weapons, has been viewed as a collateral measure designed to accomplish a specific but limited purpose. Like all collateral measures, a reason for its existence as a separate measure is that it covers only one subject, one upon which agreement may be possible before it can be reached on broader and more complex issues.
>
> The purpose of non-proliferation has always been the prevention of the spread of nuclear weapons to countries which do not have their own nuclear weapons. That purpose was articulated in the Irish resolution of 1961 which called for agreement to prevent "an increase in the number of States possessing nuclear weapons" (A/RES/1665 (XVI)).
>
> I know of no reason for believing that since 1961 the inherent concept of non-proliferation has been transmuted into something else — for example, a cut-off of production of fissionable material for nuclear weapons, a freeze of strategic delivery vehicles, or nuclear disarmament[103, 104].

The USA and the UK argued further that the discrimination against NNWS which such separate treatment of the proliferation issue necessarily entailed was also legitimate. Since maintaining the distinction between NWS and NNWS was the essence of an NPT, the obligations incumbent upon the two classes of states would inevitably be different. Being inevitable, such different obligations were thus legitimate[104, 105]. They cited as a counter-example of legitimate discrimination the Partial Test Ban Treaty which, in their view, discriminated in the opposite direction by imposing "immediate and concrete inhibitions" directly on the NWS and only "theoretical" limitations upon the NNWS[104].

Fourth, the three NWS argued not only that was it intrinsically important and legitimate to negotiate an NPT as a separate collateral measure, but also that it was necessary for expediting the negotiations. They noted that an NPT was itself an extremely complicated measure, particularly given the disagreement between the USA and the USSR on indirect dissemination of nuclear weapons to NNWS through military alliances. In their view, tying an NPT to a package of related measures, and thereby bringing into the negotiations still other and perhaps even more contentious issues, would definitely delay and perhaps make impossible agreement on an NPT. To quote Ambassador Chalfont on this point:

> I scarcely need to point out to my colleagues here, many of whom have been engaged on these negotiations far longer than I have, how difficult it is to reach any—and I stress the word 'any'—agreement on disarmament among sovereign nations each responsible for its own national security; and that difficulty will be enormously increased if we try to deal with more than one subject at a time or make agreement on one conditional on agreement on others[105].

In particular, both the USA and the USSR referred to the difficult questions of control and inspection on the territory of the USSR which would be raised by tying an NPT to such arms limitation and disarmament measures as a production cut-off and reduction of nuclear weapon stocks, a freeze and subsequent reduction of strategic nuclear delivery vehicles, and a comprehensive test ban[95, 103]. Thus, the NWS urged that an NPT be pursued as expeditiously as possible and variously described attempts by the NNWS to link it to related arms limitation and disarmament measures as "unnecessary", "imprudent", "irresponsible", "overburdening", "overloading" and "complicating"[103, 106, 107].

Fifth, the NWS asserted that an NPT agreement would specifically facilitate subsequent agreements on other related arms limitation and disarmament measures[100, 105-108]. The UK put this position most strongly when it argued in May 1967 that an NPT was a necessary precondition of such measures since "the . . . United States and the Soviet Union . . . are understandably very unlikely to begin to dismantle their own armouries while the possibility of what has been called 'horizontal' proliferation still exists"[98].

Throughout this stage of the negotiations, while arguing that an NPT be negotiated as a separate collateral measure for the above reasons, the NWS also consistently conceded the position of the NNWS that an NPT was substantively related to a number of arms limitation and disarmament measures. Indeed they admitted that the viability of an NPT was, in the long term, contingent upon agreement by NWS on measures of nuclear arms limitation and disarmament. Thus they conceded a basic premise of the Low Posture Doctrine. To quote Ambassador Chalfont on this point:

I see the non-proliferation treaty as simply the first but vital element in a broad and comprehensive strategy—a strategy for arms control, for disarmament and for international security, and for the international control of nuclear energy for the uses of peace. Certainly the treaty will not last, nor will it deserve to last, if it is used simply as a device to preserve the existing order of things, to perpetuate the oligopoly of the nuclear club.

... [The nuclear Powers] cannot expect the non-nuclear Powers of the world to deny themselves the option of possessing the most powerful military weapon the world has ever seen unless they, the nuclear Powers, are prepared themselves to engage in serious and specific measures of nuclear disarmament. Many suggestions have already been made which, in my view, contribute to that aim and deserve close and serious attention ...

But, quite apart from those detailed steps towards nuclear disarmament the principle must be accepted and clearly understood that if a non-proliferation treaty is not followed by serious attempts amongst the nuclear Powers to dismantle some of their own vast nuclear armouries then the treaty will not last, however precise its language may be. There is in my mind no doubt that, if the non-nuclear Powers are to be asked to sign a binding non-proliferation treaty, it must contain the necessary provisions and machinery to ensure that the nuclear Powers too take their proper share of the balance of obligation [98].

To some degree, however, the assertion by the NWS that an NPT would facilitate agreement on related measures was contradicted by their continuing emphasis on the currently insoluble difficulties of reaching such agreements. The USA and the USSR had consistently differed in the post-war period on proposals and agendas for arms limitation and disarmament negotiations. The USA, from a position of superiority over the USSR in central strategic nuclear capabilities between 1945 and 1965, had stressed measures constraining further production of nuclear weapons and delivery vehicles followed by gradual percentage reductions, and the importance of verification requirements[109]. The USSR, in contrast, stressed constraints on the use and deployment of nuclear weapons, pressed for more drastic amounts of nuclear disarmament, and resisted many verification proposals[110]. There was thus no clear reason to expect in the 1965-67 period that agreement on a non-dissemination clause in an NPT would directly contribute to the resolution of these long-standing differences. Indeed, it was not until the onset of SALT in November 1969 that they began to be seriously addressed. Nonetheless, by insisting that a separate NPT be reached expeditiously while acknowledging that a viable NPT depended, over time, upon subsequent nuclear arms limitation and disarmament agreements, the NWS clearly implied some sense of obligation to reach some such agreements within a limited period. To quote Ambassador Chalfont again: "What is necessary is that the sincere intent of the nuclear Powers to put such measures into effect is clear beyond doubt to the non-nuclear weapon Powers when the time comes for them to sign a non-proliferation treaty"[99].

During the two years of negotiations on the NPT after submission of the 1965 US and Soviet draft treaties, the NNWS offered various criticisms and counterproposals based on the principles outlined in Resolution 2028 (XX). With respect to measures of arms limitation and disarmament, the NNWS offered a wide range of proposals suggesting that the NWS either incorporate in an NPT or closely link to it specific measures constraining and reducing their own nuclear weapon capabilities. This general position was explicitly consistent with the Low Posture Doctrine. India and Sweden presented the most carefully elaborated proposals, outlined below. However, such views were also endorsed by Brazil, Burma, Egypt, Mexico and Nigeria, and to some degree by Canada, Ethiopia, Italy and Romania.

On 4 May 1965, even before the submission of the 1965 US and Soviet draft treaties, India had introduced in the CCD a five-point programme of non-proliferation. Its elements included:

(1) An undertaking by the nuclear Powers not to transfer nuclear weapons or nuclear weapons technology to others;
(2) An undertaking not to use nuclear weapons against countries which do not possess them;
(3) An undertaking through the United Nations to safeguard the security of countries which may be threatened by Powers having a nuclear weapons capability or about to have a nuclear weapons capability;
(4) Tangible progress towards disarmament, including a comprehensive test ban treaty, a complete freeze on production of nuclear weapons and means of delivery as well as substantial reduction in the existing stocks; and
(5) An undertaking by non-nuclear Powers not to acquire or manufacture nuclear weapons[111].

During subsequent negotiations, India reiterated that, of these five points, tangible progress towards disarmament was by far the most critical.

India consistently referred to the acquisition of nuclear weapons by existing NWS as "vertical", "actual", "existing", "present", "continuing" or "real" proliferation and to the subsequent exercise by NNWS of their nuclear weapon options as "horizontal", "further", "future", "possible" or "probable" proliferation. In the Indian view, vertical proliferation causes horizontal proliferation: the continuing acquisition of nuclear weapons by existing NWS subsequently induces NNWS to exercise their nuclear weapon option because the nuclear weapons of existing NWS pose the most severe threat to the security of NNWS. This hypothesis clearly reflects India's concern with China, an existing NWS. Thus, a prohibition on horizontal proliferation *per se*, even accompanied by various positive security guarantees, would not, in India's view, necessarily enhance the security of the NNWS, as the USA and the USSR argued. Only by constraining and then reversing vertical proliferation by all NWS, including China, would horizontal proliferation be constrained. To quote Ambassador Trevedi on this point: "So far as the nonaligned nations are concerned, security is not synonymous with protection, no matter how powerful the protector or how sincere. Real security lies in the elimination of the threat rather than in offering protection after the threat has been translated into actual aggression"[112]. Therefore, India argued that an

NPT should constrain the proliferation of all nuclear weapons, and particularly vertical proliferation, rather than merely horizontal proliferation as the US and Soviet draft treaties proposed[113].

At the outset of the negotiations, India proposed a broad programme of arms limitation and disarmament measures for incorporation in an NPT. Initially, India envisaged a procedure by which an NPT would be negotiated in two stages.

> The first stage of the treaty — or call it the partial treaty like the one on nuclear tests, for example — should incorporate provisions which are the obligations of the nuclear Powers. Under this partial treaty the nuclear Powers first undertake, under a formula acceptable to the two other Power blocs, not to pass on weapons or technology to others. Secondly, they would cease all production of nuclear weapons and delivery vehicles and agree to begin a programme of reduction of their existing stocks. Thirdly, they might also agree to incorporate in this partial treaty the other measures suggested by us in our five-point programme, as they have a certain moral and psychological value.
>
> That would be the first stage of the treaty, or a partial non-proliferation treaty. After this treaty had come into force and steps had been taken by the nuclear Powers to stop all production and embark on reduction of stocks, there would be the second stage of the treaty or the comprehensive treaty, which would provide for an undertaking by non-nuclear powers not to acquire or manufacture nuclear weapons[113].

Later in 1965, India indicated that it was flexible on the time-phasing involved in negotiating an NPT and would not require the NWS to undertake prior measures of arms limitation and disarmament. Instead, so long as undertakings by the NWS and NNWS constituted "a single or integrated process of obligations and actions", they could be simultaneous[114]. In 1966, India further amended its position and suggested that while certain measures of arms limitation and 'unarmament', including a cut-off in the production of nuclear weapons and their delivery vehicles, must be incorporated in an NPT, other measures such as reductions in stockpiles of nuclear weapons and delivery vehicles could be negotiated after the conclusion of an NPT[112]. Finally, in 1967 India proposed that an NPT specifically incorporate only a cut-off in the production of nuclear weapons. In addition, a comprehensive test ban and a freeze on all nuclear delivery vehicles were to be negotiated as collateral measures simultaneously with agreement on an NPT "in an integral and coexistent pattern"[115].

Throughout this stage of the negotiations, India consistently criticized the 1965 US and Soviet draft treaties for omitting such measures of arms limitation and disarmament, not only because such measures were a necessary condition for increasing the security of NNWS and thereby minimizing horizontal proliferation, but also because their omission from an NPT was discriminatory. India argued that the most important of the five principles enunciated in UN Resolution 2028 (XX) was operative subparagraph 2(b) which stipulated that "the treaty should embody an acceptable balance of mutual responsibilities and obligations of the nuclear and non-nuclear Powers". Moreover, this principle was the only one which stipulated that provisions for its fulfilment were to be embodied in the treaty itself. Thus, India argued that in order to meet this standard of balanced obligations and to avoid being discriminatory, an NPT must incorporate two provisions in operative articles of the treaty itself and not in some "pious preambular platitude"[112, 115, 116]. First, it must prohibit all states, not just NNWS, from producing nuclear weapons. Second, it must impose a legal obligation on the NWS to undertake subsequent limitations on and reductions of their stockpiles of nuclear weapons and delivery vehicles.

Sweden, while sharing India's views about the dangers of vertical proliferation and the discriminatory character of the 1965 US and Soviet draft treaties, consistently argued throughout this stage of the negotiations for an alternative programme. Sweden held that a comprehensive test ban and a cut-off of production of fissile material for weapon purposes in conjunction with a non-dissemination agreement would constitute both a more efficient and a more balanced package of non-proliferation measures than the approach of the USA and USSR.

First, like India, Sweden argued that its proposal would constrain horizontal proliferation more efficiently than the 1965 US and Soviet draft treaties. The three measures were inextricably linked since they raised many of the same technical questions such as, *inter alia*, the design of efficient and equitable safeguards and the control of peaceful nuclear explosives. For these reasons, Sweden proposed their simultaneous negotiation and suggested 1 January 1967 as a target date for agreement on a CTB and 1 July 1967 as a target date for agreement on a cut-off. Sweden further suggested that the entry into force of an NPT be made contingent upon these other two agreements[117]. Rather than delaying progress on an NPT, Sweden argued that simultaneous negotiation of these three interrelated measures would be efficient in that it would provide "a real grid system for incessant crosschecking of our arguments and their consequences" and place each issue in political perspective[108].

Second, like India, Sweden also argued that its package would be more balanced than a treaty limited to political pledges of non-dissemination and non-acquisition of nuclear weapons since it would constrain activities of both NWS and NNWS. Specifically, a CTB would constrain qualitative improvements in the inventories of existing NWS as well as prevent NNWS from testing explosive devices. A cut-off agreement would impose a quantitative freeze on materials available for the weapon inventories of existing NWS as well as limit NNWS to peaceful uses of fissile materials. To quote Ambassador Myrdal on this point:

> When it is sometimes said . . . that non-aligned States are claiming such measures as a kind of "price" for their adherence to a non-proliferation treaty, that is an unwarranted simplification.
>
> The true explanation . . . [is]: while we are definitely in favor of international agreements to hinder proliferation of nuclear weapons, we are also scared by the prospect that the present armament race may be allowed to continue at its perilous

pace. Therefore measures which are broad enough to bring a halt to that steeply-spiralling race at the same time as preventing additional countries from "going nuclear" are definitely to be preferred to measures that would merely achieve one and perhaps not the most far-reaching of those goals[118].

By May 1967, Sweden conceded that simultaneous acceptance of such a balanced package would not be forthcoming. Sweden continued, however, to "press for recognition of the necessity for rapid progress towards effective freezing and reversal of the present situation in the nuclear armament field". Without committing itself to a separate article on nuclear disarmament in an NPT, Sweden argued further that "any offer to forgo a nuclear option on the part of the nuclear-weapon States would serve as a reassurance to the have-nots that this first treaty would be effective and viable, and that it would truly initiate the process of nuclear disarmament"[119].

The other non-aligned NNWS at the ENDC endorsed to a greater or lesser degree the proposals of India and/or Sweden. In particular, several states such as Mexico, Burma and Ethiopia argued that the NWS must reach a comprehensive test ban in conjunction with an NPT in order to constrain the qualitative arms race[120, 121]. First, they noted that the NWS parties had undertaken to do so in the preamble of the Partial Test Ban Treaty. Second, they viewed a CTB as a minimal condition of, or a first step towards, meeting the requirement of balanced obligations under an NPT. For the NNWS, a CTB thus became perhaps the single measure of arms limitation most symbolic of the NWS' counterdiscriminatory obligations under an NPT.

There was also a continuing interest in the cut-off of production of fissile material for nuclear weapon purposes, as well as a production cut-off of nuclear weapons themselves. On 1 August 1967 Italy formally proposed a partial cut-off of the production of fissile material by means of a requirement that NWS periodically transmit some agreed amount of fissile material intended for the production of nuclear weapons to NNWS parties for peaceful purposes. Italy proposed that such a measure might be incorporated in, linked with or independent of an NPT. In any procedural arrangement, however, it would both constrain the nuclear capabilities of the NWS and promote balanced obligations in connection with an NPT[122].

Perhaps the major point at issue among the NNWS was whether an NPT should incorporate or otherwise be made contingent upon such related arms limitations and disarmament measures or whether, less stringently, the NWS should be obligated merely to undertake such measures after conclusion of an NPT[105, 106, 120, 121, 123-125]. Lack of agreement on this point is reflected in the language of the joint memoranda drafted by the eight non-aligned states in the ENDC in 1965 and 1966, in which they express their conviction that an NPT "should *be coupled with or followed by* tangible steps to halt the nuclear arms race and to limit, reduce and eliminate the stocks of nuclear weapons and the means of their delivery", and that "such steps could be *embodied in a treaty as part of its provisions or as declaration of intention*"[126, 127].

In any event, by early 1967 there was widespread recognition by the NNWS that the NWS were not willing to incorporate any related arms limitation agreements in an NPT or to make an NPT contingent upon such agreements. The non-aligned NNWS as well as Canada, Italy and Romania then converged upon a proposal originally made by Ambassador Khallaf of the UAR on 3 March 1966.

> A treaty on non-dissemination should contain a separate article under which the nuclear Powers would assume the legal obligation to halt the nuclear arms race, limit, reduce and eliminate stocks of nuclear weapons and delivery vehicles, and to that end continue and expedite negotiations in order to reach agreement on suitable concrete measures.
>
> The inclusion of such a clause in the treaty, and its application in good faith, would make it possible to assess objectively the exercise of the right of withdrawal from the treaty for nonobservance, as we have conceived it. Likewise it would solemnly confirm that the present factual nuclear monopoly will not become a legal one as a result of the non-dissemination treaty, as a substantial sector of world public opinion fears it will[106].

This proposal, a modification of which ultimately became Article VI in the final NPT, was viewed by its NNWS supporters as an explicit, legally binding obligation upon the NWS parties to agree on subsequent measures of arms limitation and disarmament as an early sequel to an NPT[106, 121, 128-131]. The NNWS viewed such a declaration of intent as an inferior substitute to incorporating within an NPT or making an NPT contingent upon specific related measures of arms limitation and disarmament. In this weaker variant, however, they still accepted the basic premises of the Low Posture Doctrine: namely, that the substantive linkage of an NPT to such subsequent measures of arms limitation and disarmament is inextricable and that early fulfilment by the NWS of such counterdiscriminatory obligations is required if an NPT is to remain viable. In effect, the conclusion of an NPT would 'buy' a limited time period within which nuclear arms limitation and disarmament negotiations were to proceed and agreements reached. To quote Ambassador Burns of Canada on this point:

> There is one prediction about this treaty which, in the Canadian view, can be made with assurance; it is that if there is no progress towards real disarmament an agreement on non-proliferation will not endure for more than relatively few years. This, we believe, is the reality of the situation, and it is not in our view highly important exactly how the obligations in respect to further measures of disarmament are formulated in the treaty which is drafted[98].

The 1967 identical draft treaties

On 24 August 1967, at the outset of the second stage of negotiations, the USA and the USSR submitted identical draft treaties to the ENDC which superseded their earlier draft treaties of 1965[132, 133]. They represented agreement between the USA and the USSR on Articles I and II of an NPT concerning the non-dissemination by NWS and the non-acquisition by NNWS of nuclear weapons. The omission of a text for Article III reflected their continuing differences concerning a system of international safeguards and in particular the conflict between the IAEA and the Euratom safeguards systems.

One important additional constraint on the NNWS was introduced in Articles I and II of the 1967 draft treaties. The 1965 US and Soviet draft treaties had referred only to controlling the dissemination and acquisition of a nuclear weapon capability. However, by 1966 the NWS stated their additional opposition to the dissemination and acquisition of capabilities for peaceful nuclear explosives (PNEs). Thus, the 1967 identical draft treaties and subsequent drafts prohibited the dissemination and acquisition of all nuclear explosive devices, regardless of whether their intended function was military or peaceful. In order to compensate for such constraints and to deter the independent development of capabilities for PNEs, the NWS undertook in the preamble of the 1967 identical draft treaties and in Article V of the subsequent drafts to provide NNWS parties to an NPT with the benefits of PNEs, were they to become economically and technologically feasible. The NWS defended this position, in which many but not all important NNWS concurred, on the grounds that the independent capability to detonate PNEs brings in train a capability to detonate nuclear weapons. Thus, the difference in the two capabilities remains merely a function of intent. Given this inextricably dual nature of nuclear explosive capabilities, the NWS concluded that omitting constraints on PNEs would constitute an unacceptable loophole in an NPT which, if exploited by a NNWS, would subvert the objective of an NPT[118, 129, 131, 134-140].

On measures of arms limitation and disarmament, however, the 1967 draft treaties clearly represented the shared views of the NWS that an NPT constituted a separate collateral measure and should exclude the related measures of interest to the NNWS. Thus, just as in the 1965 versions and consistent with the High Posture Doctrine, the 1967 draft treaties omitted all references to related arms limitation and disarmament measures in the operative body of the treaty. The 1967 draft treaties referred to such measures only in a statement of intention in the ninth preambular paragraph and in a statement of general principles in the eleventh preambular paragraph. These paragraphs read:

> The Parties to the Treaty...declaring their intention to achieve at the earliest possible date the cessation of the nuclear arms race,
>
> ...Desiring to further the easing of international tension and the strengthening of trust between States in order to facilitate the cessation of the manufacture of nuclear weapons, the liquidation of all their existing stockpiles, and the elimination from national arsenals of nuclear weapons and the means of their delivery pursuant to a Treaty on general and complete disarmament under strict and effective international control,...

The USA, the USSR and the UK argued as they had earlier that, given outstanding differences on specific measures, explicitly linking them to an NPT would jeopardize the conclusion of an NPT.

In response, and in a position consistent with the Low Posture Doctrine, the NNWS continued to advocate some form of enforceable linkage between an NPT and the process of arms limitation and nuclear disarmament. They argued that an NPT was only one of a series of measures designed to achieve the objective of nuclear disarmament and sought in some way to draft a treaty which would serve as a "basic document from which those measures would flow"[141 - 145]. While several states acknowledged that the statement of intent in the ninth preambular paragraph, the statement of general principles concerning arms limitation and disarmament measures in the eleventh preambular paragraph, and the provision for a Review Conference in Article V of the 1967 draft treaties represented some improvement in comparison with the 1965 versions, they found them collectively to be insufficient assurance that the NWS would continue the process of nuclear disarmament after an NPT had been reached[142, 145-147]. Thus they proposed both procedural and substantive amendments.

First, almost all the NNWS argued that whatever provisions relating to the arms limitation and disarmament obligations of the NWS were ultimately included in an NPT, they should take the form of a legally binding obligation in the operative body of the treaty rather than merely a statement of intention or a declaration of general principles in the preamble[146]. Ambassador Khallaf of the UAR traced the evolution of this consensus in expressing his support for the Mexican amendment which ultimately became the basis for Article VI of the final NPT.

Throughout our negotiations the link between a non-proliferation treaty and nuclear disarmament has been referred to and discussed and its importance recognized by us all. But when it comes to giving the nature, scope and form of this link a precise and acceptable definition, differences of opinion begin to appear and suggested formulas multiply.

The most radical formula would include in the non-proliferation treaty certain substantial and specific measures of nuclear disarmament. In contrast to this formula the nuclear-weapon Powers were content, in regard to their original drafts, to express the hope that the conclusion of the treaty would facilitate a start of the process of nuclear disarmament and thus excuse us from making any mention of such a link in the treaty.

However, the non-nuclear-weapon Powers asked for something more solid, more specific than this act of faith pure and simple, so the nuclear-weapon Powers are endeavouring in their new text to meet the preoccupations of the non-nuclear-weapon States and proposing a declaration of intention on nuclear disarmament. To strengthen this declaration of intention they advocate the convening of a conference of the parties to the treaty five years after its entry into force . . .

It is true that in this formulation the hope has a time limit. That already is an advance on the original drafts, which did not contemplate any link between nuclear disarmament and the treaty except a psychological link based on a mental assumption that things would move in the desired direction. In other words, in the new text we have a certain organic link between nuclear disarmament and the treaty. But is that sufficient? Ought not the declaration of intention to be made firmer becoming a distinct provision in the body of the treaty?

That was the line which the delegation of the United Arab Republic took in its statement of 3 March 1966, and which the delegation of Mexico has taken in the wording which it suggests for such an article in document ENDC/196. The merit of that formula is that it gives nuclear disarmament from the outset a more solid and more specific base: the process will thus rest on a legal base which has the merit of reflecting the general feeling of the international community[147].

Thus, the NNWS sought to offset to some degree their own binding, far-reaching and precise legal obligations not to receive or manufacture nuclear weapons, subject to international controls. Since, in the pungent words of Ambassador Trevedi of India, "There is no balance . . . between a platitude on the one hand and a prohibition on the other", the NNWS sought to introduce some such balance by formulating nuclear disarmament as a legal obligation of the NWS in an NPT[142, 148].

While agreed on the necessity to place such an obligation in the operative body of the treaty, the NNWS diverged on how the substantive obligations of the NWS to pursue a continuing process of nuclear disarmament should be formulated. As in the earlier stage of negotiations, India continued to press for the incorporation in an NPT of a cut-off of production of nuclear weapons. India suggested that the prohibition of the manufacture of nuclear weapons in Article II and the associated international controls to be elaborated in the forthcoming Article III be reformulated to apply to all states, not merely to NNWS as in the 1967 draft treaties. India submitted no formal amendments to this effect, and was the only state which maintained this position. Other NNWS no longer viewed such suggestions for incorporating related arms limitation and disarmament measures in an NPT as practicable, given the outstanding differences between the NWS on such measures[148].

Instead, most NNWS advocated incorporating in the operative body of an NPT an obligation incumbent upon NWS parties to continue the process of nuclear disarmament. Romania proposed the more stringent formulation of this obligation as amended Article IIIA which stated:

1. The nuclear-weapon States Parties to this Treaty *undertake to adopt specific measures* to bring about as soon as possible the cessation of the manufacture of nuclear weapons and the reduction and destruction of nuclear weapons and the means of their delivery.
2. If five years after the entry into force of this Treaty such measures have not been adopted, the Parties shall consider the situation created and decide on the measures to be taken.[149].

India and Burma each made similar proposals[141, 142, 148]. This more stringent formulation imposed on the NWS an obligation actually to undertake specific measures of arms limitation and disarmament.

Mexico submitted the more moderate formulation, which became the ultimate basis for Article VI in the final NPT, as amended Article IV. It stated:

Each nuclear-weapon State Party to this Treaty *undertakes to pursue negotiations in good faith*, with all speed and perseverance, to arrive at further agreements regarding the prohibition of all nuclear weapons tests, the cessation of the manufacture of nuclear weapons, the liquidation of all their existing stockpiles, the elimination from national arsenals of nuclear weapons and the means of their delivery, as well as to reach agreement on a Treaty on General and Complete Disarmament under strict and effective international control[150].

This amendment incorporated in its list of measures to be negotiated those in the preamble of the 1967 draft treaties, added the cessation of nuclear weapon tests, and clarified the wording so that agreement on such measures would not be "entirely conditional upon [their] conclusion within the framework of a treaty on general and complete disarmament"[142, 144, 146]. Brazil proposed a similar amendment[151]. This moderate formulation was endorsed by the UAR, Ethiopia and Sweden[145, 147, 148]. Canada stated it did not oppose it[148]. In contrast with that or Romania, this more moderate formulation obligated the NWS only to undertake negotiations on arms limitation and disarmament measures, rather than to undertake to adopt the measures themselves. Ambassador Castañeda justified this distinction as follows:

We are fully conscious of the obvious limits to the obligations which the nuclear Powers can assume in this respect in the present treaty. We are well aware, as we said in an earlier statement that to stipulate that the non-proliferation treaty should include specific disarmament measures to be implemented by the nuclear Powers in the immediate future, would be tantamount to opposing the very existence of a non-proliferation treaty. This fact is obvious and needs no proof or further comment. But recognition of this fact, recognition of the very limited scope of the obligations that the major Powers can assume under this treaty, is perfectly reconcilable with the desire that such obligations should be formulated more clearly and precisely, without extending their scope.

In short, the nuclear Powers cannot actually undertake to conclude further disarmament agreements among themselves; but they certainly can undertake to endeavor to do so: that is, they can certainly undertake to initiate and pursue negotiations in good faith in order to conclude such agreements. . . . Doubtless it would be an imperfect obligation, since it would not be accompanied by sanctions, but it would be more than a statement of intention. It would be a solemn recognition of the special responsibility of the nuclear Powers to adopt and implement a programme for the early reduction and possible elimination of nuclear weapons[146].

In commenting on the Mexican amendment, Canada questioned whether there was any significant difference between a legal obligation to negotiate in good faith for a certain purpose as Mexico proposed and a preambular statement of intent to achieve that purpose as the USA and USSR had proposed[148]. However, most NNWS clearly preferred to treat a continuing process of negotiation in good faith on nuclear disarmament as an explicit legal obligation of the NWS, despite their recognition that, as of 1967, existing differences would prevent the NWS from undertaking to agree on such measures.

Lastly, on a question which the NWS had introduced in the 1967 identical draft treaties, Brazil and India—unlike other NNWS in the ENDC—

specifically rejected the provisions depriving NNWS of the right to detonate PNEs under an NPT. Brazil and India consistently argued that an NPT should merely constrain the proliferation of nuclear weapons, leaving NNWS free to develop nuclear explosive devices for peaceful purposes. In their view, less economically developed NNWS should not be expected to remain technologically dependent upon advanced NWS. They asserted that such explosives had a peaceful intent and that an NPT which deprived such NNWS of their freedom to exploit all scientific, technological and economic applications of nuclear energy would be additionally discriminatory[115, 131, 142, 151-155].

Given this state of flux in the negotiations, the NWS indicated in late 1967 that they were open to suggestions and were continuing intensive negotiations among themselves on various proposed amendments[138, 141, 148, 156-158]. On 19 December 1967 the General Assembly adopted Resolution 2346A (XXII) requesting the ENDC to report to the General Assembly by 15 March 1968 on the negotiations so that the General Assembly could reconvene and hopefully consider a completed draft treaty. At the initiative of several NNWS, notably Pakistan, the General Assembly also adopted Resolution 2346B (XXII) convening a Conference of Non-Nuclear Weapon States in August-September 1968. Both meetings were designated as opportunities for the NNWS, particularly those not represented at the ENDC, to press their views concerning an NPT in the context of the comprehensive arms limitation and security régime of interest to them[80c, 159, 160].

The 1968 identical draft treaties

On 18 January 1968, at the outset of the third stage of negotiations, the USA and the USSR submitted revised identical draft treaties to the ENDC[161, 162]. With respect to measures of arms limitation and disarmament, these drafts retained in the preamble the statement of intention concerning the cessation of the arms race and the general principle of facilitating measures of arms limitation and disarmament pursuant to a treaty on general and complete disarmament (GCD). However, the January 1968 draft treaties did incorporate a new Article VI which started: "Each of the Parties to this Treaty undertakes to pursue negotiations in good faith on effective measures regarding cessation of the nuclear arms race and disarmament, and on a treaty on general and complete disarmament under strict and effective international control."

In introducing the revised treaties, both the USSR and the USA described the new articles as reflecting their desire to achieve future agreements on arms limitation and disarmament measures which would not be contingent upon their inclusion within a treaty on GCD. However, each emphasized that while Article VI obligated them to conduct negotiations on the cessation of the nuclear arms race and disarmament as well as on GCD, it did not obligate them to conclude any specific agreements. Furthermore, both states opposed listing specific measures of arms limitation and disarmament in Article VI as the Mexican amendment, upon which it was based, had done. The USA and the USSR argued, as they had since 1965, that given their outstanding differences on these measures, any effort to list them or otherwise incorporate them in the operative body of an NPT, even in terms of an obligation only to negotiate them in good faith, would delay or perhaps even jeopardize the conclusion of a treaty. To quote Ambassador Fisher of the USA on this NWS interpretation:

> From the recommendations which other members of the Committee have made on this subject, it is clear that our desire, the desire of all of us, to facilitate and not complicate subsequent nuclear disarmament negotiations is widely shared. As Mr. Castaneda has pointed out, although the nuclear Powers cannot actually undertake to conclude particular future disarmament agreements among themselves at this stage, they can undertake to initiate and pursue negotiations in good faith in order to conclude such agreements. That is essentially the content which has been given to the obligation which we are recommending be incorporated into the body of the treaty[136, 153, 163, 165, 166].

Ambassador Mulley of the UK went somewhat further than the USA and the USSR in stressing the significance of the obligation imposed by Article VI. He described it as "certainly the most important by-product of the treaty and one of its most important provisions" and noted its urgency:

> It is our desire that these negotiations should begin as soon as possible and should produce speedy and successful results. There is no excuse now for allowing a long delay to follow the signing of this treaty, as happened after the partial test ban treaty . . . before further measures can be agreed and implemented[167].

In response, the NNWS sought to strengthen these counterdiscriminatory obligations relating to arms limitation and disarmament in both the preamble and the operative body of the treaty. Their criticisms and counterproposals related to three attributes of such undertakings. First, to some degree they argued that an NPT should further specify the measures of arms limitation and disarmament which the NWS would be obligated to pursue. Second, the NNWS sought to introduce some time-urgency into an NPT, either in the form of a deadline or as a stipulation that such measures be pursued in the near future. Third, some of the NNWS sought to impose a more binding legal obligation on the NWS.

As in the second stage of the negotiations, the most stringent criticisms and proposals were offered by India and Romania, joined, to some degree, by Brazil. India, while not introducing any new proposals, continued to press for the incorporation of a cut-off of the production of nuclear weapons in an NPT. In addition, India and Romania argued that on all three of the above attributes, the new Article VI in the January 1968 draft treaties imposed insufficiently counterdiscriminatory, enforceable legal obligations upon the NWS[153, 168]. Romania introduced its amended Article VI and stressed the necessity of incorporating in an NPT a firm legal obligation binding the NWS to adopt these specific measures of disarmament as soon as possible[169]. Furthermore, the Romanian amendment, by explicitly tying such an obligation to the first mandatory review conference five years after an NPT entered into force, placed a stringent deadline on the NWS to fulfil this obligation[164, 168, 169].

Brazil criticized the January 1968 draft treaties on two of these attributes: the lack of specificity and the lack of urgency of the undertakings relating to arms limitation and disarmament. Brazil subsequently reintroduced amendments to Article VI stating: "Each nuclear-weapon State Party to this Treaty undertakes the obligation to negotiate at the earliest possible date a Treaty for the cessation of the nuclear arms race and for the eventual reduction and elimination of the nuclear arsenals and the means of delivery of the nuclear weapons"[155, 170]. Brazil emphasized the special obligation of the NWS to negotiate these specific measures of nuclear arms limitation and disarmament at the earliest possible date by relegating to a separate paragraph the requirement that all parties, both NWS and NNWS, pursue negotiations on a GCD treaty. On the third attribute, however, Brazil was less stringent than India or Romania, proposing that the NWS undertake only to negotiate such measures rather than undertake a legal obligation actually to adopt them.

Sweden introduced more moderate criticisms and counterproposals concerning the appropriate specificity, time-urgency and legal nature of the NWS' obligation to pursue arms limitation and disarmament. In an attempt to incorporate in the January 1968 draft treaties some of the elements of the 1967 Mexican amendment, Ambassador Myrdal argued:

> If, however, we compare [the wording of Article VI in the January 1968 draft treaties] with the proposals made on this very matter last year by Mexico, Romania, and Brazil, with the support of other delegations, both non-aligned and others, we must recognize that the obligations incumbent on the nuclear-weapon States are considerably weaker in the present draft. From the Mexican proposal (article IV-C) has been deleted the notion that the negotiations shall be pursued "with all speed and perseverance" and further, the clear undertaking "to arrive at further agreements". Finally, the reference in the Mexican proposal to "the prohibition of all nuclear-weapon tests" has been omitted. A similar weakening of the text can easily be noted if one compares the present wording with the proposals made by Romania and Brazil.
>
> Despite the fact that what corresponds to my Government's position would indicate the need for a much stronger commitment without further delay to steps of effective nuclear disarmament, I am mindful of the difficulties involved. As has been stated, it would hardly be feasible in legal terms to enter into obligations to arrive at agreements. Further, to enumerate some specific measures might be counterproductive, as agreements on certain other scores may come to present opportunities for earlier implementation.
>
> For those reasons the Swedish delegation today will restrict to two its suggested amendments to article VI, both being of such a nature that they are not expected to create any difficulties in regard to the substance. We simply propose the inclusion of the words "at an early date", thus introducing once more the sense of urgency which

we all feel presses for further measures to halt the nuclear arms race. We also proposed, for the sake of making clear the main goal of these negotiations the insertion of the word "nuclear" before the word "disarmament".

If one is to be able to keep the amendment of article VI to such very modest dimensions, it practically presupposes some strengthening of the preamble. In the present wording of the preamble there are three paragraphs — the last ones — which deal with further disarmament measures. Even if it can be surmised that the comprehensive test ban is one of the measures implicit in the reference to "the cessation of the nuclear arms race", we urge that this important measure of disarmament be spelt out specifically somewhere in the treaty. This is all the more justified as we can rest this particular case on a previously-accepted commitment[155, 171-174].

Thus, in terms of further specificity, Sweden proposed that Article VI emphasize measures of nuclear disarmament and that the preamble specify an intention "to achieve" a comprehensive test ban, the issue which Sweden deemed most immediately connected with the obligations undertaken in Article VI. In terms of time-urgency, Sweden proposed that Article VI enjoin the parties to negotiate measures of arms limitation and disarmament "at an early date". Sweden envisaged this date to be within the subsequent few years, arguing that the time necessary for signing and ratifying an NPT and negotiating the associated safeguards agreements required under Article III constituted a natural testing period for the NWS' willingness to undertake serious negotiations on further arms limitation and disarmament measures. Finally, Sweden clearly accepted the formulation first put forward by Mexico in 1967 that the NWS be legally obligated to undertake only the negotiation of arms limitation and disarmament measures, not the measures themselves.

On 22 February 1968, the UK endorsed the Swedish amendments of Article VI, with minor modifications acceptable to Sweden, as well as a general preambular commitment to a CTB[136, 174]. These amendments were also endorsed by Bulgaria, Canada, Ethiopia, Mexico, the UAR and Nigeria[173, 175-177].

The March 1968 joint draft treaty

On 11 March 1968, at the outset of the fourth stage of negotiations, the USA and the USSR submitted a revised joint draft treaty[178]. With respect to measures of arms limitation and disarmament, it retained the preambular statements of intention concerning the cessation of the arms race and the general principle of facilitating measures of arms limitation and disarmament pursuant to a treaty on GCD. In addition, it included a new preambular paragraph, similar to that proposed by Sweden, which read: "Recalling the determination expressed by the Parties to the Partial Test Ban Treaty of 1963 in its preamble to seek to achieve the discontinuance of all test explosions of nuclear weapons for all time and to continue negotiations to this end". Originally, Sweden had proposed a paragraph which would recall the determination of parties to the Partial Test Ban Treaty "to achieve" a CTB[171]. Both the USA and the USSR explained that the language in the March 1968 draft treaty recalling the determination "to seek to achieve" a CTB replicated the exact wording of the preamble in the Partial Test Ban Treaty and as such, was more authentic than the Swedish amendment[168]. However, this change did, in effect, moderate the objective from that of achieving a CTB to seeking through negotiations to achieve one.

The March 1968 draft treaty also incorporated in Article VI the two Swedish amendments emphasizing the urgency of the obligation and the goal of nuclear disarmament, so as to read:

> Each of the Parties to this Treaty undertakes to pursue negotiations in good faith on effective measures relating to cessation of the nuclear arms race at an early date and to nuclear disarmament, and on a Treaty on general and complete disarmament under strict and effective international control.

This version of Article VI was subsequently incorporated in the final NPT.

Both the USA and the USSR defended these arms limitation and disarmament provisions in the ENDC and in the General Assembly debate as the maximum possible agreement currently achievable within an NPT and dismissed pressure for more binding and substantial undertakings as jeopardizing the conclusion of an NPT. They argued further that those states which criticized these provisions as overly general and non-enforceable would have to be satisfied with them since the control of proliferation through an NPT was a necessary pre-condition for further progress towards nuclear arms limitation and disarmament[179-182].

On the other hand, both the USA and the USSR underscored the seriousness with which they interpreted these provisions. Ambassador Kuznetsov of the USSR described Article VI as an unprecedented commitment and asserted that the USSR would be "most responsible" in attempting to fulfil it[181]. Ambassador Goldberg of the USA described it with comparable emphasis as a meaningful undertaking and asserted that "the permanent viability of this treaty will depend in large measure on our success in the further negotiations contemplated in article VI"[183]. Furthermore, each major NWS repeatedly described an NPT as directed towards several major objectives: stopping the proliferation of nuclear weapons to additional nation-states; sharing the benefits of peaceful uses of nuclear energy; strengthening the security of all signatories, particularly the NNWS; and, most significant in this connection, contributing to the cessation of the nuclear arms race and to disarmament[181, 183-185]. Thus, both the USA and the USSR clearly acknowledged the direct linkage between an NPT and the progress of their subsequent negotiations on nuclear arms limitation and disarmament. Thus, at least in their declarations, they conceded the logic of the Low Posture Doctrine. In this connection, on 1 May 1968, the USA, the USSR, the UK and 17 other states introduced draft resolution A/C.1/L.421 endorsing the March 1968 draft treaty and requesting "the ENDC urgently to pursue negotiations of effective measures relating to the cessation of the nuclear arms race at an early date and to nuclear disarmament, and on a treaty on general and complete disarmament under strict and effective international control".

Throughout the General Assembly session, however, at least 44 NNWS continued to make two major criticisms of the provisions for arms limitation and disarmament in the March 1968 draft treaty and suggested several specific additional proposals[180-182, 184, 186-201].

First, 23 NNWS—Afghanistan, Argentina, Austria, Brazil, Ceylon, Cyprus, Guyana, India, Indonesia, Italy, Jordan, Laos, Libya, Mauritania, Nigeria, Pakistan, Peru, Sierra Leone, South Africa, Thailand, Uganda, Yugoslavia and Zambia—criticized the arms limitation and disarmament provisions in the March 1968 draft treaty on the grounds that they imposed inadequate counterdiscriminatory obligations and responsibilities upon the NWS[7b, 181, 182, 184, 186-193, 195, 197].[1] In their view, the draft treaty thus failed to fulfil the second and third principles of Resolution 2028 (XX) requiring that the treaty "embody an acceptable balance of mutual responsibilities and obligations" incumbent upon the NWS and the NNWS, respectively, and that the treaty be "a step towards the achievement of general and complete disarmament and, more particularly, nuclear disarmament".

Second, 19 NNWS—Algeria, Argentina, Brazil, Dahomey, El Salvador, Ethiopia, Guyana, India, Jamaica, Japan, Jordan, Madagascar, Malaysia, Malta, Panama, Peru, Rwanda, South Africa, and Trinidad and Tobago—specifically criticized the provisions of Article VI[180-182, 184, 187, 189, 190, 193-196, 198, 199, 201]. They argued that the obligation in Article VI "to pursue negotiations in good faith" was vague, subjective, insufficiently compelling and merely an expression of intention to undertake negotiations in good faith which, in the previous 20 years, had not resulted in significant arms limitation and disarmament agreements. Thus it did not constitute a firm, binding legal obligation which would require the NWS to negotiate and subsequently reach agreements on specific measures of arms limitation and disarmament within a reasonably short time period. Conversely, some of these states argued that the NPT not only required that negotiations be held in good faith, but that—consistent with the requirements of the Low Posture Doctrine—it also created an expectation that successful agreements on specific arms limitation and disarmament measures would be forthcoming from such negotiations within a reasonably short time period.

Brazil and India continued to take the most stringent position on these issues. They announced their unwillingness to adhere to the evolving NPT

[1]Lloyd Jensen has made a content analysis of the 1968 General Assembly debate on the NPT, in which he found that of the 87 delegations participating, 44 per cent faulted the March 1968 draft treaty for its lack of linkage to broader disarmament measures. Thirty-six per cent also criticized it for lack of balance in the obligations relating principally to disarmament and/or the control systems which were to be imposed on the NNWS. In this analysis, Jensen's two categories have been dissected, the states specifically concerned with arms limitation and disarmament have been identified, and their particular criticisms further specified[7].

due to its discriminatory imposition of obligations and privileges and, in particular, to the absence of any enforceable obligation, subject to sanctions and a time limit, incumbent upon the NWS to proceed with nuclear arms limitation and disarmament. Indeed, both Brazil and India argued that by the omission of such an obligation, the March 1968 draft treaty implicitly authorized the continuation of vertical proliferation for at least 25 years, the initial period of its duration. To quote Ambassador Husain of India on these points:

It has been said that if we were to attempt to achieve agreement on all aspects of disarmament at this time, the negotiating difficulties would be insurmountable and we would end by achieving nothing. Agreement on all or even some aspects of disarmament, if I may say so, is not what many countries within and without the Eighteen-Nation Committee on Disarmament — and certainly not my country — have urged. . . . Furthermore, the delegation of India has never suggested that a non-proliferation treaty should in itself become a vehicle or a measure of full-fledged nuclear disarmament. But we do feel that so long as the augmentation and sophistication of nuclear weapons by the existing nuclear-weapon Powers continues unchecked the interests of the security of the world will not be advanced. Measures which do not involve an element of self-restraint on the part of all States — nuclear weapon States as well as non-nuclear-weapon States — cannot form the basis for meaningful international agreements to promote disarmament.

. . .

Article VI does not give any tangible form to the declaration of good intent, there being no sense of compulsive obligation or even a sense of urgency to pursue negotiations for nuclear disarmament as a preliminary to general and complete disarmament. What is required is something in the nature of a nuclear moratorium, as was suggested in [the Fanfani Proposal of] 1965, of which the essential element was that if nuclear disarmament was not achieved within a specified time limit, the non-nuclear-weapon Powers, as an instrument of persuasion and pressure, would reserve to themselves the resumption of their freedom of action.

. . .

India, as is well known, has pleaded for various collateral disarmament measures for two decades now and has always regarded the non-proliferation treaty as one of those measures. But we still need to be convinced that the draft treaty before us does amount to a collateral disarmament measure. In order to become generally acceptable the treaty must have a provision for some degree of compulsiveness and a reasonable time limit, indicating a sense of urgency on the part of the nuclear-weapon States to move towards nuclear disarmament, thus paving the way for general and complete disarmament; otherwise, this non-proliferation treaty — and it does not matter by whom or by how many it is signed — will not be effective and will not last and our labours will have been in vain[189, 193].

Alternatively, some of these states suggested several specific measures of arms limitation on which the NWS should reach agreement in the reasonably short-term future. Fourteen NNWS — Afghanistan, Algeria, Ceylon, Cyprus, Ghana, Kenya, Liberia, Malaysia, Nepal, Pakistan, the Philippines, Sweden, Venezuela and Yugoslavia — advocated agreement on a CTB in the short-term future[180, 181, 186-188, 190-193, 195, 196, 200]. Nine states — Afghanistan, Cuba, Cyprus, India, Italy, Pakistan, the

Philippines, Sweden and Yugoslavia — advocated a cut-off of production of fissile material for nuclear weapon purposes and seven NNWS — Algeria, Cyprus, Guyana, Panama, Romania, Uganda and Venezuela — proposed a freeze on the production of nuclear weapons[181, 182, 186-188, 190-193, 199]. Finally, four NNWS — Cuba, Libya, the Philippines and Romania — called for disarmament measures, specifically reductions of nuclear weapons, delivery vehicles and/or conventional military capabilities[187, 191, 197]. It was clear that these NNWS expected some progress in the form of agreements over time on such measures of nuclear arms limitation and disarmament, and not merely negotiations on such measures, as an important *quid pro quo* for their accepting an NPT and forgoing their own nuclear weapon option.

The May 1968 joint draft treaty

To meet such objections, on 31 May 1968, the USA and the USSR submitted a revised joint draft treaty to the General Assembly which incorporated one amendment, proposed by Yugoslavia, related to arms limitation and disarmament[184, 202]. It included an additional declaration of intent in the preamble "to undertake effective measures in the direction of nuclear disarmament". The NWS and their supporters also revised the accompanying draft resolution to stress that effective *measures* of rather than steps towards nuclear arms limitation and disarmament must follow an NPT and to request in operative paragraph 4 that the NWS as well as the ENDC urgently pursue negotiations on such measures. Finally, given the continuing opposition to the treaty, the resolution was revised so that the General Assembly would merely be asked to commend the May 1968 draft treaty rather than to endorse it.

Many of the NNWS have, however, remained unpersuaded that the arms limitation and disarmament provisions of the NPT, among other provisions, are adequate. Of the 44 NNWS which criticized the NPT during the General Assembly debate on the grounds that it imposed insufficient arms limitation and disarmament obligations on the NWS, 14 — Algeria, Argentina, Brazil, Cuba, Guyana, India, Indonesia, Mauritania, Pakistan, South Africa, Sri Lanka, Trinidad and Tobago, Uganda and Zambia — have not ratified the NPT as of June 1976. Thus 32 per cent of the NNWS objecting to the arms limitation and disarmament provisions of the NPT have remained outside the treaty, as compared with 29 per cent of the 1968 membership of the United Nations which participated in the debate.

The resolution commending the final May 1968 version of the NPT was adopted by the General Assembly on 12 June 1968 as Resolution 2373 (XXII) by a vote of 95 to 4 with 21 abstentions. Among those states opposing or abstaining were 11 NNWS — Algeria, Argentina, Brazil, Burma, Cuba, India, Mauritania, Rwanda, Sierra Leone, Uganda and Zambia — which had criticized the arms limitation and disarmament provisions of the NPT in either the ENDC or the General Assembly. Only two of these states have now ratified the NPT.

II. Security guarantees

The second major issue dividing the NWS from the NNWS concerned what types of security guarantee should be linked to an NPT and in what manner. As noted in chapter 2, security guarantees can vary along a number of dimensions. They can be extended in a formal treaty or in some less legally binding or tacit form. They can be provided unilaterally by a single guarantor state, multilaterally by a group of guarantor states, or jointly through an international organization. They can be extended unconditionally or contingent upon specified conditions to specified addressees. Finally, as discussed in the previous chapter, they can either be positive or affirmative, in the sense of guaranteeing the security of a particular state or group of states against attack or the threat of attack, or they can be negative, in the sense of guaranteeing not to attack or threaten attack against a particular state or group of states.

Both types of security guarantees are conceivably relevant to an NPT, since any means by which a guarantor state can contribute to the security of an addressee state may provide an incentive to the addressee to forgo its nuclear weapon option. However, since positive security guarantees were presumably already operative in some degree for NNWS allied to NWS, the major issue in the negotiations on an NPT primarily became that of providing some form of security guarantees to non-aligned NNWS. During negotiations on an NPT, various NNWS advocated both types of guarantee. First, many NNWS sought different forms of a negative security guarantee from NWS parties to an NPT, either in the NPT itself or in some related undertaking, not to use or threaten to use nuclear weapons against certain NNWS. Second, several NNWS sought positive security guarantees from NWS parties to an NPT, either in the NPT itself or in some related undertaking, to protect them against attack or threatened attack by other NWS, particularly if such other NWS were not constrained by having previously subscribed to a negative security guarantee[7c, 117]. This composite position of the NNWS was consistent with the security requirements of the modified Low Posture Doctrine.

The 1965 draft treaties

Neither the US nor the Soviet draft treaty of 1965 included provisions for any type of security guarantee. This omission was consistent with the NWS position that an NPT should be strictly limited and not incorporate measures other than the non-dissemination by NWS and the non-acquisition by NNWS of nuclear weapons. Neither the USA nor the USSR ever formally incorporated in their various draft treaties provisions for any type of security guarantee except a commitment to respect the principle of denuclearized zones. However, by early 1966 both the USA and the USSR had recognized the need to provide security guarantees to non-aligned NNWS through some undertaking in connection with an NPT, in order to give them incentives to forgo their nuclear weapon option.

The Western and Eastern states in the ENDC, however, initially proposed quite different types of assurance, consistent with their long-standing differences concerning constraints on the use of nuclear weapons. The West, including the USA, the UK, Canada and Italy, consistently supported positive security guarantees. The Eastern states, including the USSR, Bulgaria, Czechoslovakia, Poland and Romania, consistently supported negative security guarantees in general and what became known as the 'Kosygin Proposal' (see below) in particular.

The basic Western position pre-dates the beginning of sustained negotiations on an NPT. Immediately after China first tested a nuclear explosive device on 16 October 1964, the USA announced that "if [NNWS] need our strong support against some threat of nuclear blackmail, then they shall have it"[203]. In a message to the ENDC on 27 January 1966, US President Johnson reaffirmed this pledge:

So that those who forswear nuclear weapons may forever refrain without fear from entering the nuclear arms race, let us strive to strengthen United Nations and other international security arrangements. Meanwhile, the nations that do not seek the nuclear path can be sure that they will have our strong support against threats of nuclear blackmail[109].

This unilateral declaration of intent offered positive security guarantees to NNWS which felt their security threatened by China's nuclear capability. It became the standard US response to demands by NNWS for security guarantees. At the same time, the USA and its military allies criticized negative security guarantees in general and the Kosygin Proposal in particular as unenforceable, unverifiable, disruptive of the current state of mutual deterrence, and ineffective in the face of threats emanating from NWS such as China which would not become parties to an NPT[99, 204, 205]. This US position on security guarantees is entirely consistent with the High Posture Doctrine.

Soviet Premier Kosygin first elaborated the Eastern position in a message to the ENDC on 1 February 1966 when he proposed a conditional negative security guarantee which became known as the Kosygin Proposal [206]. Although never formally introducing it as an amendment to its 1965 draft treaty, the USSR subsequently suggested that the text for such a clause might be: "The parties to the treaty possessing nuclear weapons undertake not to use nuclear weapons and not to threaten the use of such weapons against States which do not possess nuclear weapons and in whose territory, territorial waters and air space there are no foreign nuclear weapons"[207]. This conditional negative security guarantee would be extended to NNWS which, first, were parties to an NPT and, second, had no nuclear weapons on their territory. These conditions would exclude from the guarantee's coverage certain military allies of NWS — among them most notably FR Germany — which permitted the stationing of nuclear weapons on their territories[106]. It would thus create two categories of NNWS under an NPT: those legally exempt from nuclear targeting in the event of nuclear war and those against whom such targeting was permissible.

The Kosygin Proposal adapted to an NPT two long-standing Soviet disarmament proposals: an international convention prohibiting the use of nuclear weapons, and an agreement prohibiting the foreign deployment of nuclear weapons. Both proposals challenged US nuclear policy which traditionally justified foreign deployment, use, and in certain instances first use of nuclear weapons as means of deterring attack or waging a defensive war[208]. At the same time, the Eastern states argued that it was a negative security guarantee which would best satisfy the perceived security needs of the non-aligned NNWS rather than the unilateral positive security guarantee offered by the USA[106, 209]. The Eastern states were intensely critical of the Western states for their traditional lack of interest in proposals constraining the use of nuclear weapons and for their refusal to negotiate seriously on such proposals, including the Kosygin Proposal[134]. Thus the Soviet position was, at this juncture of the negotiations, consistent with the Low Posture Doctrine.

No conclusive resolution of this issue emerged in 1966 and 1967, but there was clear movement towards a compromise between the Western and Eastern positions. First, the UK as well as Canada and Italy sought to shift the initiative to the non-aligned NNWS. They argued that the Western and Eastern approaches dealt with different perceived security needs and suggested that, since the NWS were each willing to offer some form of guarantee in connection with an NPT, the non-aligned NNWS toward whom the two approaches were directed should inform the ENDC about which type of guarantee they preferred[99, 104, 121, 210].

Second, in a position closer to that of the non-aligned NNWS, Ambassador Burns of Canada suggested that an NPT might reaffirm in a general article,

> the principle that nuclear Powers were responsible for ensuring against nuclear attack or threats of it the safety of non-aligned nations which agreed to abstain from acquiring nuclear weapons. Such an article could serve as a basis for separate bilateral or multilateral agreements to be made between those non-nuclear nations which felt that they needed guarantees and one or more of the nuclear Powers[106].

Canada also suggested that an NPT might incorporate the Kosygin Proposal as a fairly simple article, thereby incorporating some form of both positive and negative security guarantees[106, 211].

Third, as the Western and Eastern states moved towards agreement on the questions of NNWS access to or control of nuclear weapons through military alliances and the European option, the USSR deferred to the continuing US opposition to negative security guarantees in general and the Kosygin Proposal in particular and quietly dropped this proposal[7d, 154, 208]. Thus, by early 1967, negotiations on how to link some form of positive security guarantees to an NPT were taking place between the NWS outside the formal ENDC meetings. Negative security guarantees were no longer under consideration by either NWS. Publicly, the USA and the USSR continued to defend their original position that agreement on an NPT should not be delayed by efforts to incorporate in it contentious measures such as specific forms of security guarantees.

In contrast to the position of the NWS expressed in their 1965 draft treaties, the non-aligned NNWS advocated various forms of negative security guarantee applicable to broad categories of states, both to protect the security of NNWS and to impose counterdiscriminatory obligations upon the NWS. On 31 August 1965, Nigeria introduced the most elaborate proposal incorporating negative security guarantees and a no-first-use undertaking. It suggested, first, that the NWS should "neither indulge in nuclear blackmail of smaller states nor threaten their sovereignty with conventional arms", and, second, that NWS forswear the use of nuclear weapons or, at least, "give a categorical assurance that they will in no circumstances whatever use nuclear weapons against non-nuclear States and that even against each other they will not be the first to use them"[97]. It was only such guarantees of abstention from the use or threatened use of nuclear weapons by NWS against all NNWS, pending ultimate nuclear disarmament divesting NWS of their capacity to use nuclear weapons, that would protect NNWS from nuclear blackmail. During this period, Nigeria consistently argued that such negative security guarantees against nuclear attack and nuclear blackmail were critical if an NPT were to protect the security of NNWS and thereby create incentives for relevant NNWS to forgo permanently their nuclear weapon option[125, 212].

On 16 March 1967, the UAR proposed that an NPT incorporate an article obligating the NWS not to use or threaten to use nuclear weapons against NNWS in order to impose balanced obligations upon the NWS. To quote Ambassador Khallaf on this point:

> It is inconceivable that the non-nuclear States which under the treaty would renounce nuclear weapons would quite simply agree by the same act to reserve to nuclear Powers the privilege of threatening them or attacking them with those same weapons.
>
> ...A non-proliferation treaty which excluded the obligation of the nuclear Powers not to use or threaten to use nuclear weapons against non-nuclear States would not only 'enshrine' the monopoly of nuclear attack of the nuclear Powers but would also — and this is a serious matter — increase the striking power of the existing nuclear members. This would in no way correspond either to the spirit or to the purpose of a non-proliferation treaty[213].

Two broader constraints on the use of nuclear weapons were endorsed by certain NNWS as contributing to an international security régime which, by minimizing the use of nuclear weapons in general, would promote the security of the NNWS, impose counterdiscriminatory obligations on the NWS, and thus reduce incentives for NNWS to exercise their nuclear weapon option. First, Brazil as well as Nigeria proposed a commitment by NWS to forgo the first use of nuclear weapons against each other[97, 214]. Second and most broadly, Ethiopia — consistent with its sponsorship of UN Resolution 1653 (XVI) (a Declaration on the Prohibition of the Use of Nuclear and Thermo-nuclear Weapons) — proposed a total prohibition of the use or threatened use of nuclear weapons by NWS[105, 215].

In a related position, Mexico, Ethiopia, Sweden, the UAR and Romania endorsed the principle of denuclearized or nuclear-free zones in various regions, particularly Latin America and Africa[123, 204, 215-218]. Such zones would provide both for the non-acquisition of nuclear weapons by NNWS and for the non-deployment and non-use of nuclear weapons by the NWS in the affected regions. The Mexican representative, as Chairman of the Preparatory Commission for the Treaty for the Prohibition of Nuclear Weapons in Latin America — which was opened for signature in February 1967 — became the chief spokesman for the NNWS on nuclear-free zones. Mexico argued that it was absolutely necessary that an NPT incorporate a commitment to the principle of denuclearized zones which, as noted previously, were a combined non-proliferation agreement, non-deployment zone and negative security guarantee. In pursuit of this objective, on 19 July 1966 Mexico proposed that subparagraph 2(e) of UN Resolution 2028 (XX) be directly incorporated into an NPT as the following article: "No provision of this treaty shall be interpreted as in any way lessening or restricting the right of any group of States to conclude regional treaties in order to ensure the total absence of nuclear weapons in their respective territories"[216].

Finally, for various reasons, a number of non-aligned NNWS endorsed the Kosygin Proposal. The UAR did so as "an application in a specific case of the general rule of total prohibition" of nuclear weapons, a general position which the UAR had always supported[106]. Mexico endorsed the proposal with the interpretation that it undertook to respect regional denuclearized zones, and thus would serve as an incentive to reach such agreements[120, 216]. Several NNWS endorsed the Kosygin Proposal simply because it would incorporate at least a contingent negative security guarantee in an NPT. During the 1965 General Assembly, Nigeria had strongly criticized the unwillingness of the USA and the USSR to include in Resolution 2028 (XX) some reference to "adequate guarantees by the nuclear Powers not to use nuclear weapons against non-nuclear Powers under any circumstances whatsoever, or to threaten to use them"[219]. Nigeria, India and other NNWS thus endorsed the Kosygin Proposal because it was to some degree responsive to their position that negative security guarantees would both enhance the security of the NNWS and impose some counterdiscriminatory obligations upon the NWS[112, 219].

In pursuit of these objectives, the NNWS sponsored Resolution 2153 (XXI) at the 1966 General Assembly calling for the prompt conclusion of an NPT incorporating two types of multilateral negative security guarantee. First, it urged all NWS to refrain from the use or threat of use of nuclear weapons against states parties to treaties establishing denuclearized zones. Second and more generally, it asked the ENDC to consider the Kosygin Proposal "and any other proposals that have been or may be made for the solution of this problem". Just as in the case of arms limitation and disarmament measures, the NNWS viewed the incorporation of various forms of negative security guarantee in an NPT as a major component of the arms limitation and security régime of interest to them. This position was entirely consistent with the Low Posture Doctrine's stress upon deployment and use constraints and inconsistent with the tenets of the High Posture Doctrine. To quote Ambassador Gomez Robledo of Mexico speaking in February 1966 on this point:

> This question of balance is identical with, or at any rate implies or presupposes, the other question of the so-called guarantees which the non-nuclear Powers must unquestionably receive if they are to take such a serious step as that of limiting their sovereignty and restricting their security[120].

Most non-aligned NNWS coupled their support for these various forms of negative security guarantee with opposition to unilateral positive security guarantees from particular NWS. Many feared that such guarantees might infringe upon their non-alignment and result in undesirable tacit alliance with one or another NWS[121, 220]. The UAR, with obvious reference to the Middle East, argued that the provision of positive security guarantees by a particular NWS to one of a pair or group of hostile NNWS might encourage this trend, since it would "tempt other nuclear Powers to offer the same guarantee and thus the effect of a nuclear guarantee would be, in the end, to place the world in a situation where vast areas were divided under a nuclear trusteeship of this or that Power"[123]. Other NNWS, particularly Brazil and India, simply doubted whether unilateral positive security guarantees could ever be made credible for non-aligned states unprotected by military alliances[142, 214].

Nigeria, India and, to some degree, Canada did, however, advocate multilateral or joint positive security guarantees, a position consistent with the modified Low Posture Doctrine. In 1965 Nigeria proposed the development of sufficient UN peacekeeping capabilities to "safeguard and guarantee the territorial integrity of States"[97]. In May 1965 India, as noted above, similarly proposed an "undertaking through the United Nations to safeguard the security of countries which may be threatened by Powers having a nuclear weapons capability or about to have a nuclear weapons capability"[111]. Throughout this stage of the negotiations, India was the strongest advocate of multilateral positive security guarantees, just as it had been the strongest advocate of a broad programme of arms limitation and disarmament measures as an integral part of an NPT. Both positions stemmed from India's peculiar position as a NNWS. First, it was a non-aligned country not linked to any NWS through either an alliance or some sort of vague 'nuclear umbrella'. Second, it had an advanced nuclear industrial capacity. Third, it perceived a continuing threat of nuclear attack or blackmail from a neighbouring NWS, China[221a]. Thus, India felt it needed multilateral undertakings from the USA, the UK and the USSR to ensure its security from the threat of China.

By 1967, however, India had become convinced that the NWS would not provide credible and effective multilateral positive security guarantees to non-aligned NNWS within an NPT régime, just as they would not undertake substantial arms limitation and disarmament measures. Furthermore, any negative security guarantees as well as arms limitation and disarmament measures incorporated in or related to an NPT would be irrelevant for India's security needs, at least in the short term, since, by 1967, India was publicly dismissing as a "utopian dream" the possibility that China would ratify an NPT[221b]. Thus, by 1967 it was clear that the Chinese threat, rather than any provisions which a forthcoming NPT might incorporate, would determine India's final attitude towards maintaining its own nuclear weapon option and hence towards signing an NPT. To quote Ambassador Trevedi, speaking on the peculiar difficulties of India's case, "the question of security is a much wider issue and is relevant irrespective of a treaty on non-proliferation of nuclear weapons"[115]. The position of India during this stage of negotiations points up perhaps most clearly the failure of the evolving NPT to provide a comprehensive arms limitation and security régime which would protect the interests of all important NNWS.

The 1967 identical draft treaties

At the outset of the second stage of the negotiations, the USA and the USSR responded to the concerns of Mexico and other NNWS by including in the 1967 draft treaties a twelfth preambular paragraph: "Noting that nothing in this Treaty affects the right of any group of States to conclude regional treaties in order to assure the total absence of nuclear weapons in their respective territories..." Otherwise, they omitted references to security guarantees. In explaining the draft treaties at the ENDC and the UN, the USA and the USSR referred to their private negotiations on some form of security gurantees to be offered in conjunction with an NPT. Both states made it clear, however, that security guarantees would not be incorporated in an NPT[157, 222-224].

For their part, the NNWS continued to be critical of the virtual omission of security guarantees from the 1967 draft treaties and urged the NWS to provide them[148, 225]. They welcomed the preambular declaration that an NPT would not interfere with the establishment of regional denuclearized zones. However, Mexico proposed that the exact language of this declaration of principle become an amended Article IV-B in the operative body of the treaty, since it was "an authentic legal provision" with "obvious legal effects"[146, 150]. Brazil proposed a similar amendment[151].

Several NNWS proposed amendments which incorporated three other types of security guarantee in the operative body of an NPT. First, Romania proposed as amended Article III-B a broad negative security guarantee to be extended to all NNWS party to an NPT: "Nuclear-weapon States Parties to this Treaty solemnly undertake never in any circumstances to use or threaten to use nuclear weapons against non-nuclear weapon States which undertake not to manufacture or acquire nuclear weapons" [149]. Romania advocated this amendment on three grounds. First, Romania had consistently supported measures requiring the non-utilization of all weapons of mass destruction and this provision was merely a limited application of that principle. Second, a provision requiring the non-utilization of nuclear weapons against NNWS parties to an NPT would both ensure their security and impose a counterdiscriminatory obligation upon the NWS parties. Third, a provision requiring the non-utilization of nuclear weapons against NNWS would, by limiting the utility of nuclear weapons, facilitate nuclear disarmament[142, 226]. The Romanian position was endorsed by Burma and Switzerland, in a memorandum to the ENDC[141, 227].

Second, the UAR suggested that the Kosygin Proposal be incorporated in an NPT as amended Article IV-A: "Each nuclear-weapon State undertakes not to use, or threaten to use, nuclear weapons against any non-nuclear weapon State Party to this Treaty which has no nuclear weapons on its territory"[228]. The UAR argued that since this conditional formulation of a negative security guarantee had been supported by the General Assembly in Resolution 2153 (XXI), it already enjoyed legitimacy and was thus a realistic basis for negotiation[143].

Third, Nigeria, in a shift of emphasis, proposed that a positive security guarantee be incorporated in an NPT as Article II-A: "Each nuclear weapon State Party to this Treaty undertakes, if requested, to come to the aid of any non-nuclear weapon State which is threatened or attacked with nuclear weapons"[229]. Ambassador Sule Kolo justified this amendment on the grounds that NNWS parties needed a 'nuclear umbrella' pending the elimination of nuclear weapons[226].

Hence, during this stage of the negotiations, it became clear that many of the NNWS, with the exceptions of Nigeria and India, were still primarily interested in some form of negative security guarantee. Furthermore while advocating somewhat different substantive security guarantees, most of the NNWS agreed that such assurances should be incorporated as legal obligations of the NWS in the operative body of an NPT, both because security guarantees were central to the security needs of the NNWS and because such guarantees would impose counterdiscriminatory obligations on the NWS[143, 226, 230].

Canada, however, opposed all three proposed security guarantees, as did the NWS, on this procedural point. While emphasizing the importance of meeting the perceived security needs of all NNWS, both allied and non-aligned, and advocating both positive and negative security guarantees, Canada argued that attempting to incorporate variegated and complex security guarantees in an NPT would unduly delay its conclusion[156, 224, 231]. Alternatively, Ambassador Burns suggested that all such assurances be provided through one of two procedures outside an NPT:

> The first would be by means of unilateral declarations to be made by nuclear Powers at the time the treaty is opened for signature. Separate declarations using similar language might record the intention of nuclear Powers to assist non-nuclear States which sign the non-proliferation treaty and which are subsequently subjected to nuclear attack or threatened with it. They might also incorporate an undertaking that nuclear weapons will not be used against non-nuclear States signatory to the treaty which are not allied with a nuclear Power.
>
> A second method of achieving the same general objective might be to proceed by way of a United Nations resolution incorporating in its substantive paragraphs assurances similar to those I have just mentioned. Such a resolution might also take account of the special responsibility placed on the Security Council under the United Nations Charter for maintaining peace and resisting aggression[224].

Despite the contrary views of many NNWS on this point, this procedural solution was ultimately proposed by the NWS and adopted in Security Council Resolution 255 and its associated unilateral declarations.

The 1968 identical draft treaties

At the outset of the third stage of the negotiations, the USA and the USSR accepted the Mexican proposal and incorporated in the January 1968 draft treaties a new article VII providing that: "Nothing in this Treaty affects the right of any group of States to conclude regional treaties in order to assure the total absence of nuclear weapons in their respective territories"[175]. This concession was consistent with the Low Posture Doctrine's support of nuclear-free zones. Otherwise, the USA and the USSR omitted references to broader security guarantees. They confirmed that they were continuing private negotiations on this issue and reiterated their position that such guarantees should be offered through the United Nations in conjunction with but separate from an NPT[163, 165]. They continued to oppose the inclusion of such guarantees in an NPT and argued, as Canada had, that the interests of both allied and non-aligned NNWS were too complicated and variegated to be reduced to credible treaty provisions[163, 165].

The NNWS, for their part, continued to object to this omission. Several NNWS reiterated their earlier proposals that various security guarantees be incorporated in an NPT as a counterdiscriminatory obligation incumbent upon the NWS. The UAR again proposed that the Kosygin Proposal be included as a treaty provision[176]. Romania and Ethiopia advocated incorporating in an NPT broader negative security guarantees for all NNWS or for all NNWS parties to an NPT[164, 173]. Such proposals were, however, specifically rejected by the USA[165]. On the other hand, in a memorandum transmitted to the ENDC, FR Germany advocated some form of positive security guarantee as a specific treaty provision rather than as a separate declaration in conjunction with an NPT. It urged that such a guarantee undertake to protect NNWS from "political threats, political pressure, or political blackmail" by NWS[232]. Nigeria also resubmitted its earlier proposal incorporating a positive security guarantee in an NPT[233]. To a substantial degree, however, discussion of security guarantees by the NNWS became quite perfunctory during this stage of the negotiations, as they awaited a joint US-Soviet proposal on the issue.

The 1968 tripartite draft Security Council resolution and March joint draft treaty

On 7 March 1968, the USA, the UK and the USSR introduced their tripartite draft Security Council resolution on security guarantees in the ENDC and described their forthcoming unilateral declarations of intention upon which the resolution rested[234, 235]. Four days later, on 11 March 1968, the USA and the USSR submitted a revised joint draft treaty which, with the exception of Article VII committing the signatories to the principle of denuclearized zones, included no provisions for security guarantees. On 14 March 1968 the ENDC submitted its summary report on the negotiations to the UN General Assembly[236-238]. This report attached as annexes both the US-Soviet March 1968 draft treaty and the US-British-Soviet draft Security Council resolution, as well as a list of proposals submitted by the other ENDC member states during 1967-68 so as to indicate that the documents submitted by the NWS were not necessarily supported by other ENDC member states. The General Assembly reconvened on 24 April 1968 to conclude the final negotiations on an NPT. Thus, during this period, the draft resolution concerning security guarantees and the associated unilateral declarations of intention were drafted, introduced, criticized and passed as an adjunct to an NPT, and they can be analysed as an integral part of the debate on the NPT.

The package of joint and unilateral security guarantees introduced on 7 March 1968 outside of, but in conjunction with, the NPT constituted the response of the NWS to the continuing demands of the NNWS for various types of security guarantees. The draft resolution read:

The Security Council,

1. Recognizes that aggression with nuclear weapons or the threat of such aggression against a non-nuclear State would create a situation in which the Security Council, and above all its nuclear-weapon State permanent members, would have to act immediately in accordance with their obligations under the United Nations Charter;

2. Welcomes the intention expressed by certain States that they will provide or support immediate assistance, in accordance with the Charter, to any non-nuclear-weapon state Party to the Treaty on the Non-Proliferation of Nuclear Weapons that is a victim of an act or an object of a threat of aggression in which nuclear weapons are used;

3. Reaffirms in particular the inherent right, recognized under Article 51 of the Charter, of individual and collective self-defense if an armed attack occurs against a Member of the United Nations, until the Security Council has taken measures necessary to maintain international peace and security.

The three related unilateral declarations of intention submitted by the USA, the UK and the USSR and referred to in paragraph 2 of the draft resolution were identical. To quote the operative sections of the US declaration:

Aggression with nuclear weapons, or the threat of such aggression, against a non-nuclear-weapon State would create a qualitatively new situation in which the nuclear-weapon States which are permanent members of the United Nations Security Council would have to act immediately through the Security Council to take the measures necessary to counter such aggression or to remove the threat of aggression in accordance with the United Nations Charter, which calls for taking "effective collective measures for the prevention and removal of threats to the peace, and for the suppression of acts of aggression or other breaches of the peace". Therefore, any State which commits aggression accompanied by the use of nuclear weapons or which threatens such aggression must be aware that its actions are to be countered effectively by measures to be taken in accordance with the United Nations Charter to suppress the aggression or remove the threat of aggression.

The United States affirms its intention, as a permanent member of the United Nations Security Council, to seek immediate Security Council action to provide assistance, in accordance with the Charter, to any non-nuclear-weapon State party to the treaty on the non-proliferation of nuclear weapons that is a victim of an act of aggression or an object of a threat of aggression in which nuclear weapons are used.

The United States reaffirms in particular the inherent right, recognized under Article 51 of the Charter, of individual and collective self-defence if an armed attack, including a nuclear attack, occurs against a Member of the United Nations, until the Security Council has taken measures necessary to maintain international peace and security.

The United States vote for the draft resolution before us and this statement of the way in which the United States intends to act in accordance with the Charter of the United Nations are based upon the fact that the draft resolution is supported by other permanent members of the Security Council which are nuclear-weapon States and are also proposing to sign the treaty on the non-proliferation of nuclear weapons, and that these States have made similar statements as to the way in which they intend to act in accordance with the Charter.

These undertakings were thus designed solely to provide positive assurances of assistance to NNWS parties to an NPT either jointly through the Security Council in case of aggression or the threat of aggression with nuclear weapons by a NWS, or unilaterally or severally under Article 51 in case of armed attack with nuclear weapons.

Twenty-two states — namely, the three sponsoring NWS, 13 NATO or WTO members, Argentina, Ethiopia, Finland, Ireland, Paraguay and Senegal — supported the draft resolution and the associated unilateral declarations in the General Assembly and Security Council debates. They did so on four grounds. First, while admitting that such assurances did not provide an absolute guarantee of assistance against nuclear attack, they argued that the draft resolution and the associated unilateral declarations constituted a significant political development which went some distance towards safeguarding the security of the NNWS[180, 181, 183]. Specifically, these undertakings represented a degree of political collaboration between the USA and the USSR which gave unprecedented evidence of reduced tension and increased cooperation between them. Such collaboration would both help maintain the peace and to some degree render credible and effective a joint deterrent against aggression or threatened aggression with nuclear weapons. This position was explicitly consistent with the tenets of the High Posture Doctrine in its primary emphasis upon positive security guarantees issued by the two powers. To quote US Ambassador Goldberg on this point:

It is no secret that these Powers command the overwhelming preponderance of nuclear-weapon power in the world today. That these major nuclear Powers, whatever their respective views on other matters, have now united in this proposal is a political fact of the first order. It means that they consider that their respective vital national interests demand that there shall be no nuclear aggression, and no threat of nuclear aggression, from any quarter; and that those countries that forgo nuclear weapons by adhering to the non-proliferation treaty should not thereby feel any loss of security.

Thus, the sponsorship of this Security Council draft resolution by these three nuclear-weapon Powers introduces a powerful element of deterrence against nuclear aggression or the threat of such aggression. It also is a fact of history, as we who represent our countries in the United Nations well know, that where the three nuclear nations which have developed this Security Council draft resolution have joined in support of a proposed action by the Council, such action has usually been forthcoming and effective [180].

Second, such undertakings by three NWS to agree in advance to come to the assistance of NNWS through the UN system of collective security in the event of aggression or the threat of aggression with nuclear weapons provided a kind of international nuclear deterrent which did not jeopardize the non-aligned status of many NNWS[200]. Third, such undertakings constituted a counterdiscriminatory obligation of the NWS which would introduce additional balance into an NPT[190, 239]. Fourth, such undertakings cited the right of individual and collective self-defence under Article 51 of the UN Charter, thereby both leaving intact the positive security guarantees already extant through existing military alliances, and legitimizing other unilateral or multilateral actions which the three NWS might take outside of the Security Council[51b, 182].

Of the 19 NNWS which supported the draft resolution and the associated unilateral declarations, only two — Argentina and Turkey — have failed to ratify the treaty. Thus only 11 per cent of the NNWS supporting the security guarantees proposed by the NWS have remained outside the NPT, as compared with 29 per cent of the 1968 membership of the United Nations which participated in the debate.

The draft resolution and the associated unilateral declarations were opposed in the General Assembly and the Security Council debates by many NNWS on two major grounds: the exclusion of negative security guarantees and the inadequacy of the positive security guarantees which were included.

Twenty-five NNWS criticized the NWS for not providing any negative security guarantees other than the commitment to the principle of denuclearized zones incorporated in Article VII of the NPT[180, 182, 184, 186, 187, 189, 192-194, 196, 197, 199-201, 240-243]. During the debate, this group of states suggested six different substantive forms of constraints on the use of nuclear weapons, ranging from limited, contingent norms of prohibition to unconditional ones. Their shared position was entirely consistent with the requirements of the Low Posture Doctrine.

First, and most limited in scope, Rwanda argued that Article VII should be strengthened so that NWS parties would undertake within the operative body of the NPT not to attack states within a denuclearized zone[196].

Second, three NNWS — Iran, Tanzania and Yugoslavia — supported the Kosygin Proposal, arguing both that it had been previously supported in the General Assembly Resolution 2153 (XXI) and that it would merely extend to all NNWS parties without nuclear weapons on their territories the same guarantees which the USA and the UK had already extended to Latin American states under Additional Protocol II of the Treaty of Tlatelolco [193, 196, 242].

Third, 11 NNWS — Afghanistan, Argentina, Brazil, Cyprus, Dahomey, Guyana, Libya, Pakistan, Romania, Trinidad and Tobago, and Uganda — called for the negative security guarantee previously proposed by Romania in the ENDC: namely, that NWS parties to an NPT undertake not to attack or to threaten to attack NNWS parties with nuclear weapons[180, 182, 186, 187, 189, 191, 192, 197, 201]. This request for an assurance that NWS parties would not use nuclear weapons against NNWS parties to an NPT was defended as a direct and rather minimal means of promoting balanced, counterdiscriminatory obligations in an NPT.

Fourth, seven NNWS — Greece, Israel, Jamaica, Kenya, Liberia, Nepal and Panama — advocated a negative security guarantee through which NWS parties would undertake not to attack or to threaten to attack with nuclear weapons any NNWS regardless of whether such NNWS was party to the NPT[180, 184, 187, 192, 197, 199, 200].

Fifth, Burma called for a more general prohibition against the first use of nuclear weapons in any contingency[240].

Sixth and most unconditionally, eight NNWS — Afghanistan, Albania, Ethiopia, Guyana, Nepal, Trinidad and Tobago, Uganda and Yugoslavia — expressed support for a non-contingent prohibition on the use of nuclear weapons against NWS and NNWS alike, consistent with General Assembly Resolution 1653 (XVI)[182, 186, 189, 191, 193, 194,200, 201].

In addition, 13 of these NNWS argued specifically that their preferred use constraint be incorporated in the operative body of the treaty as a binding legal obligation of the signatory NWS, both to enhance the security of the NNWS and to provide additional balance in an NPT[180, 182, 186, 191 - 193, 196, 199, 242].

Of the 25 NNWS which criticized the NPT for not containing various use constraints, 10 — Albania, Argentina, Brazil, Burma, Guyana, Israel, Pakistan, Tanzania, Trinidad and Tobago, and Uganda — have failed to ratify the treaty. Thus 40 per cent of the NNWS objecting to the absence of negative security guarantees or other use constraints in the NPT have remained outside the Treaty, as compared with 29 per cent of the 1968 membership of the UN.

Forty-one NNWS criticized the draft Security Council resolution and the associated unilateral declarations during the debates in the General Assembly and the Security Council on the second major ground that they constituted inadequate positive security guarantees[180 - 182, 184, 186, 187, 189 - 193, 195 - 201, 242, 244]. These states criticized both the general procedural inadequacies of these security provisions and their failure to meet the specific security needs or political objectives of particular NNWS. A few of these NNWS proposed, in addition, certain means of strengthening the positive security guarantees to be associated with the final NPT.

In terms of their general procedural inadequacies, 30 of these NNWS criticized the draft Security Council resolution and the associated unilateral declarations for their reliance upon the institutional structures and procedures of the United Nations to protect the security interests of the NNWS. These states — namely, Afghanistan, Brazil, Ceylon, Chile, Colombia, Cyprus, Dahomey, El Salvador, Ghana, Greece, Indonesia, Iran, Jamaica, Jordan, Kenya, Lesotho, Madagascar, Malta, Mauritania, Nepal, Pakistan, Panama, Rwanda, Sierra Leone, Spain, Tanzania, Thailand, Uganda, Yugoslavia and Zambia — emphasized three reasons why the proffered security provisions constituted at best uncertain and at worst potentially meaningless positive security guarantees.

First, at least 12 of these NNWS argued that the draft resolution and its associated declarations were merely vague statements of intention which suggested no new procedures and imposed no additional obligations upon the sponsoring NWS beyond those already undertaken through the UN Charter [186, 189, 191, 193, 196, 199, 242, 244]. To quote Tanzanian Ambassador Danieli on this point:

> The draft resolution contains no promise or obligation; it merely recognizes and reaffirms the obvious — namely, that aggression would require the Security Council to take appropriate action, and that all States have an inherent right of self-defense in the event of armed attack. The draft resolution also contains a reference to intentions expressed elsewhere by "certain States" to assist non-nuclear-weapon States. Needless to say, such vague references create no juridical obligations[242].

Thus, these security provisions, by creating no new counterdiscriminatory obligations, provide neither additional security for NNWS parties nor additional balance for the NPT.

Second, 17 NNWS specifically noted that taking action against such aggression through the Security Council would be subject to the veto of the five permanent members of the Security Council, four of which were then NWS[180, 186, 187, 190-193, 195, 196, 198, 242, 244]. Thus, if the positive security guarantees were to be exercised, they would have to win the support or at least the acquiescence of all five permanent members. Many of these NNWS noted that the permanent members had opposed each other in past crises threatening the maintenance of international peace and security such as Berlin, the Congo, the Dominican Republic, South Africa, the Middle East and Viet-Nam, when these issues had been brought before the Security Council. Upon occasion, they had either exercised or threatened to exercise their veto on behalf of their own national interests. In particular, several African states — including Dahomey, Ghana, Kenya, Madagascar, Mauritania, Rwanda, Sierra Leone, Tanzania, Uganda and Zambia — cited the recent threatened use of its veto by the UK in opposition to the imposition of stringent sanctions against Southern Rhodesia[180, 187, 191, 193, 195, 196, 198, 242]. Thus, in a much more pressing crisis involving aggression with nuclear weapons or the threat of such aggression, these NNWS argued that there would be even less reason to expect consensus and common interests among the five permanent members of the Security Council. To quote Ambassador Zollner of Dahomey on this point:

> We are well aware that any move to act by the Security Council can be blocked by a single veto issued by any one of its permanent members. We know that the veto power is not merely a formality, and that it has been employed on many occasions in the past in far less serious circumstances; we now that there have been even more occasions when the mere certainty that one or other of the permanent members would use it has nipped in the bud many proposals and desires for action which were none the less necessary. If the first paragraph of the resolution is not a useless repetition of the Charter, what is its meaning? Does it mean that in specific cases such as this nuclear Powers that are permanent members of the Council might begin to consider some curbs on their veto power, which is primarily responsible for the impotence of the Security Council? We would be happy to learn that, but in the meanwhile we feel there is room for reasonable doubt[180].

Third, 15 NNWS noted that the event bringing the proposed security guarantees into action was aggression with nuclear weapons or the threat of such aggression[180, 182, 184, 186, 187, 190-192, 195-198, 200, 241]. Thus, these guarantees turned on the concept of aggression, the definition of which had long been a subject of dispute in the United Nations. To quote Ghanaian Ambassador Akwei on this point:

> The sponsors of the proposed resolution on security guarantees to be submitted in the Security Council undertake to go to the assistance of non-nuclear-weapon States which are victims of an act or an object of a threat of aggression. But the record of the United Nations is replete with instances of inaction on the part of the Security

Council precisely because the permanent members of that organ have not been able to identify the same State as the aggressor. Without wishing to appear unduly pessimistic, it seems to my delegation, however, that given the prevailing international atmosphere and conflicting concepts of the defence of the national interest, and the treaty ties binding different sets of allies, it would not be an exaggeration to say that we are very far from a generally acceptable standard of what constitutes aggression. Therefore, until we all agree on what constitutes aggression, how are the sponsors of this security guarantee to determine who has been the aggressor in certain circumstances?[195]

In order to mitigate this definitional problem, Afghanistan, Dahomey, Iran and Pakistan argued that the resolution should be amended so as to apply to the use or threatened use of nuclear weapons rather than applying to the disputed concept of aggression[180, 186, 196, 241]. Alternatively, Zambia proposed that aggression be specifically defined "within the context of and for inclusion in the treaty" and Cyprus urged that the NWS agree on a general definition of aggression which, *inter alia*, could be applied under this resolution[187, 192].

In addition to these general procedural inadequacies, 16 of these NNWS criticized the draft Security Council resolution and the associated unilateral declarations for their failure to meet specific security needs or political objectives of particular NNWS. These states — namely, Albania, Algeria, Cuba, Dahomey, Ghana, India, Indonesia, Israel, Jordan, Mauritania, Nepal, Pakistan, Sierra Leone, Tanzania, Uganda and Zambia — mounted four specific arguments.

First, six states — Algeria, Cuba, India, Israel, Pakistan and Zambia — opposed the draft resolution on the grounds that operative paragraph 2 limited the proffered guarantees to parties to an NPT rather than extending them to all NNWS members of the United Nations, irrespective of whether they were parties to this particular treaty[187, 191, 192, 241]. These states argued that such a discriminatory restriction in effect denied to non-parties their right to receive assistance against aggression which was their due under the UN Charter. To quote Indian Ambassador Parthasarathi on this point:

I should like to emphasize that any security assurances that might be offered by nuclear-weapons States could not and should not be regarded as a *quid pro quo* for the signature of a non-proliferation treaty. A non-proliferation treaty should be judged by itself and on its own merits. As I have already stated, the threat of nuclear weapons to non-nuclear-weapons States arises directly from the possession of such weapons by certain States. That threat has existed in the past and will continue to remain, even after a non-proliferation treaty has been concluded, until such time as the nuclear menace has been eliminated altogether. The assurance of security to non-nuclear-weapon States is an obligation on the nuclear-weapon States, and not something which they could or should offer in return for the signature by non-nuclear-weapon States of a non-proliferation treaty.

The basis for any action by the Security Council for the maintenance of international peace and security is the Charter of the United Nations. Any linking of security assurances to the signature of a non-proliferation treaty would be contrary to its provisions, because the Charter does not discriminate between those who might adhere to a particular treaty and those who might not do so. Under Article 24 of the Charter, the Members of the United Nations have conferred on the Security Council the primary responsibility for the maintenance of international peace and security, and have agreed that, in carrying out its duties under this responsibility, the Security Council acts on their behalf. Article 24 then goes on to say that in discharging its duties the Council shall act in accordance with the purposes and principles of the United Nations. The purposes and principles are contained in Articles 1 and 2 of the Charter. One of the cardinal principles is that of sovereign equality, that is, the equality of rights and benefits under the Charter for all Members of the United Nations. The second, and equally important, principle is that all Members shall fulfil in good faith the obligations assumed by them in accordance with the Charter. It would thus be clear that, while the permanent members of the Security Council have a special obligation and responsibility for the maintenance of international peace and security, they are precluded from adopting a discriminatory approach in situations involving the security of States, including that arising from the threat or the use of nuclear weapons against non-nuclear-weapon States[241].

Second, Pakistan argued that the inherent right of individual and collective self-defence under Article 51 referred to in the draft resolution and the unilateral declarations applied only to armed attack with nuclear weapons, and not the threat of attack, thus further constraining the potential scope of the guarantees[241].

Third, seven states — namely, Algeria, Cuba, Dahomey, Ghana, Jordan, Tanzania and Zambia — noted that the security provisions did not cover situations in which one of the guarantors might initiate aggression with nuclear weapons or threaten such aggression against a NNWS[180, 182, 187, 191, 195, 241, 242]. Nor did they relate to the initiation of nuclear war by one of the NWS guarantors against another NWS. In these cases, each of the guarantors, as a permanent member of the Security Council, would obviously protect itself by vetoing any proposed collective action against it. Thus, states which felt threatened, or were politically sympathetic to states which felt threatened, by one of the three NWS guarantors could derive no additional security assurances whatsoever from the draft Security Council resolution. Furthermore, all NNWS as potential 'innocent bystanders' in the event of a nuclear exchange involving a guarantor NWS, would remain unprotected. To quote Zambian Ambassador Mapanza on these points:

Aggression is aggression, whether it comes from China, France, the United Kingdom or the United States. In this treaty the drafters have anticipated only aggression by China or France, the two nuclear Powers which are not likely to sign the treaty. What if one of the three nuclear States party to the treaty launched a nuclear attack on a non-nuclear State party to the treaty? My delegation would like to know what would be the role of the United Nations in the event one of the three nuclear States party to the treaty launched an attack, using nuclear weapons, on a non-nuclear State.

Furthermore, my delegation would like to know what would be the role of this Organization if two nuclear States party to the treaty were to embark on a nuclear war, the fall-out of which would invariably affect citizens of non-nuclear States not engaged in the conflict. These questions give rise to serious concern to my delegation. Yes, the possibility of this happening exists. Is this Organization going to resign itself to the role of passive onlooker, paralysed either by the veto or by big-Power self-advantage? This kind of treaty is, to say the least, unsatisfactory and if we are going to renounce our right to self-defence — and this is what this treaty amounts to for non-nuclear States — then we must have meaningful and workable guarantees. As the treaty stands this requirement has not been met[187].

In sum, these guarantees would be ineffectual against any NWS which was a permanent member of the Security Council. Although not a guarantor, France could utilize its veto to protect itself in the event of its initiating aggression with nuclear weapons or threatening such aggression. Therefore, given the existence of five NWS in 1968, the joint security guarantees could only be activated against an existing NWS as long as PR China, the remaining NWS, did not occupy the permanent seat in the Security Council reserved for China. When PR China did take this seat in 1971, the joint security guarantee adopted in 1968 became meaningless against existing NWS, since all were then protected by the veto. However, given the concurrence of all five permanent members of the Security Council, the joint security guarantee remained theoretically applicable to states which might in future exercise their nuclear option, as India did in 1974.

Fourth, four states — Albania, Algeria, Nepal and Zambia — specifically objected to what they viewed as the prejudicial treatment of China in being singled out as the only object of the joint security guarantees[181, 187, 189, 200]. Albania and Tanzania, clearly representing the position of China, argued further that these security provisions constituted part of an attempt by the USA and the USSR to establish a condominium in international politics, in this instance by specifically threatening China[189, 242]. All four of the states which ultimately voted against General Assembly Resolution 2373 (XXII) commending the NPT — Albania, Cuba, Tanzania and Zambia — were sympathetic to the position of China on the treaty.

More generally, at least nine NNWS — including Algeria, Ghana, Indonesia, Mauritania, Nepal, Sierra Leone, Tanzania, Uganda and Zambia — argued that to be effective in the long term, joint positive security guarantees would have to take into account all existing NWS, including France and China which had taken no part in devising the draft Security Council resolution. These states argued that at a minimum China would have to be seated in the UN and particularly in the Security Council before effective security guarantees could be designed[187, 191 - 193, 195, 200, 241].

In addition to these procedural and substantive criticisms of the draft Security Council resolution and the associated unilateral declarations, seven NNWS — Ceylon, Ghana, Kenya, Sierra Leone, the Sudan, Trinidad and Tobago and the UAR — proposed strengthening the positive security guarantees to be associated with an NPT[187, 190, 192, 195, 196, 199, 201]. Several of these states, concerned about hostile neighbours exercising their nuclear weapon option in the future, proposed that such guarantees be made more timely and automatic than the current Security Council procedures would permit and that they be bolstered by increased UN peacekeeping capabilities.

Finally, 11 NNWS — namely Afghanistan, Barbados, Colombia, Dahomey, El Salvador, Kenya, Mauritania, Panama, Rwanda, Sierra Leone and Tanzania — specifically proposed that the positive security guarantees proffered in the tripartite draft Security Council resolution be incorporated in the operative body of the treaty as a legally binding, counterdiscriminatory obligation incumbent upon the NWS parties[180, 186, 187, 193, 196-199, 242]. Such a positive security guarantee within the NPT would not only introduce additional balance into the treaty but would also obviate the possible exercise of a veto in the Security Council. To quote Ambassador Zollner of Dahomey on this point:

> Our first objection has to do with the form in which these guarantees are given. We feel that security guarantees play a part in establishing the "acceptable balance of mutual responsibilities and obligations of the nuclear and non-nuclear Powers" which resolution 2028 (XX) requires of the treaty. The provisions dealing with security guarantees should therefore appear in the body of the treaty, and should constitute multilateral obligations having the same force as those assumed by the non-nuclear States under the treaty. We see no reason why these guarantees should take the form of simple resolutions or unilateral declarations. Of course, we are happy to note the excellent intentions which the nuclear Powers propose to express in their various declarations, some of which are very important and significant. However, if such declarations can suffice, why should we not be satisfied with similar statements and mere declarations from the non-nuclear countries and why would a treaty be necessary at all?[180]

The May 1968 joint draft treaty

In response to these criticisms and proposals, the USA and the USSR incorporated a new preambular paragraph in the draft treaty of 31 May 1968, which emphasized the connection between an NPT and the obligations of states under the UN Charter not to use or threaten to use force. The paragraph, which had been proposed by Mexico and Japan, read:

> Recalling that, in accordance with the Charter of the United Nations, States must refrain in their international relations from the threat or use of force against the territorial integrity or political independence of any state, or in any manner inconsistent with the purposes of the United Nations, and that the establishment and maintenance of international peace and security are to be promoted with the least diversion for armaments of the world's human and economic resources[186, 190].

The NWS argued that reaffirming the principles underlying the UN Charter in the NPT would reinforce the related obligations they would undertake in the tripartite draft Security Council resolution[184]. With this slight concession, the NWS adhered to a position on security guarantees entirely consistent with the High Posture Doctrine.

The NWS made no other revisions concerning security guarantees in either the May 1968 draft treaty or the March 1968 draft Security Council resolution. Immediately after the General Assembly passed Resolution 2373 (XXII) commending the NPT on 12 June 1968, the USA, the UK and the USSR submitted the tripartite draft resolution to the Security Council. The Security Council approved it on 19 June 1968 as Resolution 255 by a vote of 10-0 with five abstentions — Algeria, Brazil, France, India and Pakistan. Of these five states, none has signed or ratified the NPT.

Moreover, many NNWS remain convinced that Security Council Resolution 255 and its associated unilateral declarations provide inadequate positive security guarantees. Of the 41 NNWS which made this general argument during the General Assembly and Security Council debates, 19 — Albania, Algeria, Barbados, Brazil, Chile, Colombia, Cuba, India, Indonesia, Israel, Mauritania, Pakistan, Spain, Sri Lanka, Tanzania, Trinidad and Tobago, the UAR, Uganda and Zambia — have not ratified the NPT. Thus 46 per cent of the NNWS objecting to the positive security guarantees related to the NPT have remained outside the treaty, as compared with 29 per cent of the 1968 membership of the UN. It has been noted above that 40 per cent of the NNWS objecting to the absence of negative security guarantees and other use constraints have remained outside the NPT. These positions of many NNWS are consistent with the logic of the modified Low Posture Doctrine, which emphasizes the significance both of negative security guarantees and other use constraints and of credible, joint and non-hegemonic positive security guarantees in an effective anti-proliferation régime.

III. Provisions for review, duration and withdrawal

The third major issue dividing the NWS from the NNWS concerned the procedural provisions for review, duration and withdrawal to be incorporated in an NPT lest its substantive provisions not be observed or prove inadequate over time. As it became clear that the NWS would not incorporate within an NPT the comprehensive arms limitation and security régime of interest to the NNWS, the negotiations increasingly focused upon the procedures under which an NPT would remain in force.

The 1965 draft treaties

With respect to the procedural issues of review, duration and withdrawal, both the US and the Soviet draft treaties provided that the NPT be of indefinite or unlimited duration, subject to the right of any party to withdraw from the treaty if "it decides that extraordinary events, related to the subject matter" of the treaty have jeopardized the supreme interests of its country[93, 94]. This language was identical to that in the 1963 Partial Test Ban Treaty.

In addition, the US 1965 draft treaty provided for a conference to be held upon agreement of two-thirds of the parties some specific number of years after entry into force of the treaty "in order to review the operation of the Treaty". In presenting the draft on 17 August 1965, US Ambassador Foster specifically linked the Review Conference to "the wide concern recently expressed by many participants in the discussions here and in the Disarmament Commission that a treaty such as this should be accompanied by progress to halt and reduce rising nuclear stocks"[245]. The following year, Ambassador Fisher of the USA elaborated on this connection:

> Our draft treaty...contains a review provision which is designed to permit non-nuclear weapon States to consider, after a limited period from the entry into force of the non-proliferation treaty, whether they are satisfied with the progress then made in halting the arms race.
>
> This review provision should be viewed in the light of the preambular reference to ou[r] common objective in the United States draft: "to achieve effective agreements to halt the nuclear arms race, and to reduce armaments, including particularly nuclear arsenals"[106].

In other words, the review conference was to constitute "the means of redress" for the NNWS should the NWS not fulfil their counterdiscriminatory obligations to reach nuclear arms limitation and disarmament agreements within a limited period[107].

The NNWS position on these procedural issues evolved as they concluded that the NWS were unwilling to incorporate specific measures of nuclear arms limitation and disarmament and specific security guarantees in an NPT. Thereupon, the NNWS became increasingly concerned that an NPT contain review procedures and conditions of duration and withdrawal which would forge linkages between the duration of an NPT and progress towards the related comprehensive arms limitation and security régimes of interest to them. To quote Ethiopian Ambassador Imru on this point:

> The Ethiopian delegation therefore believes that a high measure of priority should be given to a non-proliferation agreement. However, if we regard the non-proliferation measures simply as an instrument of containment and thus an end in itself, our achievement will be illusory. It will perhaps be necessary to put a time limit on its duration so that we can gauge its efficacy by progress registered in the sphere of general disarmament and related measures — for it is legitimate to ask ourselves how long a non-proliferation agreement can last if it does not lead to the limitation, reduction, and eventual elimination of nuclear weapons and remains a simple ordinance of self-denial on the part of the non-nuclear world[105, 121, 124].

This concern with issues of review, duration and withdrawal thus constituted a kind of intermediate position between the NNWS preference for incorporating specific measures of arms limitation and security guarantees in the NPT itself and the NWS preference for treating an NPT separately as at least a short-term end in itself[246].

One variant of this intermediate position was the Draft of a Unilateral Non-acquisition Declaration circulated by Italy on 14 September 1965, which came to be known as the Fanfani Proposal[247, 248]. Italy proposed that, pending conclusion of a permanent NPT, individual NNWS sign a unilateral moratorium renouncing the acquisition of nuclear weapons for a time period to be specified. Such an undertaking would, in Italy's view, bring pressure on the NWS to fulfil the specific demands of the NNWS concerning arms limitation and disarmament within the stipulated time-limit since, if they were not met, the NNWS would be freed from any obligations not to acquire nuclear weapons. In Foreign Minister Fanfani's words:

> In that way a respite would be given to the anxiety about nuclear dissemination and, moreover, a factor of pressure and persuasion would be created which could be brought to bear on the nuclear countries in order to spur them to conclude a general agreement, thus speeding up the process of nuclear disarmament[245, 247, 248].

Alternatively, such a moratorium might be extended if progress were being made towards "international agreements to prevent the spread of nuclear weapons, to halt the nuclear arms race and to reduce nuclear arsenals" [247, 249].

Reactions of non-aligned NNWS to the Fanfani Proposal were mixed. India viewed the approach as fully consistent with that of the non-aligned NNWS, in that it attempted to deal simultaneously with horizontal and vertical proliferation[246]. Sweden was more negative, arguing that the Fanfani Proposal might have a counter-productive effect in reducing the urgency of the non-dissemination issue. In Sweden's view, the moratorium should be effective only over a relatively short time period of, perhaps, two years, and only if matched by pledges from the NWS that they would do everything possible to "achieve definitive results in relation to a Comprehensive Test Ban and a Non-Proliferation Treaty"[220]. Ethiopia was openly critical, arguing that the moratorium was non-binding and thus would merely give licence to the NWS for further acquisition of nuclear weapons[120].

In the Joint Memorandum of 1966, the non-aligned NNWS in the ENDC bypassed the Fanfani Proposal in favour of instituting binding review procedures in a permanent NPT. In reiterating support for UN Resolution 2028 (XX), the memorandum pointed out that:

> Principle (d) requires that there should be workable provisions to ensure the effectiveness of the treaty. The eight delegations consider that such provisions should guarantee compliance with the obligations of the treaty. They, furthermore, believe that an essential provision to ensure the effectiveness of the treaty, not least in the context of the undertakings on further steps towards disarmament mentioned in the preceding paragraphs, would be that of making the treaty subject to periodic reviews[127].

NNWS support of periodic reviews was thus explicitly linked to their support of binding commitments incumbent upon the NWS to undertake subsequent measures of arms limitation and disarmament. Formal review procedures in an NPT were seen as the best available means of putting pressure on the NWS to fulfil some of their counterdiscriminatory obligations pertaining to the comprehensive arms limitation and security régime of interest to the NNWS[217].

The 1967 identical draft treaties

At the outset of the second stage of negotiations, the USA and the USSR, drawing on the provision in the US 1965 draft treaty, provided in Article V of the 1967 draft treaties that: "Five years after the entry into force of this Treaty, a conference of Parties to the Treaty shall be held in Geneva, Switzerland, in order to review the operation of this Treaty with a view to assuring that the purposes and provisions of the Treaty are being realized". In introducing the 1967 draft treaties, US Ambassador Foster explained the purpose of the mandatory review conference:

> This will provide an opportunity for non-nuclear and nuclear-weapon States alike to assess whether the treaty is accomplishing its primary purpose of preventing the spread of nuclear weapons, and also its purposes of easing international tensions and facilitating agreement on cessation of the nuclear arms race and on disarmament. The review conference is thus relevant to the question of further measures of disarmament, a question which has been of such interest to many members of this Committee[250].

The UK emphasized, further, that five years was a short time period after which parties would be called to account at the Review Conference for their compliance with the purposes and provisions of the treaty[251].

On 10 October 1967 the UK introduced an amendment to Article V which attempted to clarify this linkage:

> Five years after the entry into force of this Treaty, a conference of Parties to the Treaty shall be held in Geneva, Switzerland, in order to review the operation of this Treaty with a view to assuring that the *purposes of the Preamble* and the provisions of the Treaty are being realised[252].

This formulation, which was subsequently endorsed by Italy, specifically linked a Review Conference to the preambular language relating to arms limitation and disarmament[253]. It was a clear attempt on the part of a NWS to avoid incorporating a legal obligation to continue the process of disarmament in the operative body of an NPT. However, it also specifically mandated a Review Conference to monitor the progress of the NWS in achieving arms limitation and disarmament, thereby at least conceding the logic of the Low Posture Doctrine.

While providing for this single Review Conference, the 1967 draft treaties, like those of 1965, stipulated that the NPT was to be of unlimited duration. Each party had a right to withdraw from the Treaty if it decides that extraordinary events, related to the subject matter of this Treaty, have jeopardized the supreme interests of its country. It shall give notice of such withdrawal to all other Parties to the Treaty and to the United Nations Security Council three months in advance. Such notice shall include a statement of the extraordinary events it regards as having jeopardized its supreme interests.

This procedure for giving notice, by involving the Security Council, can be viewed as a means of evoking a possible positive security guarantee for the affected parties. If the party contemplating withdrawal had a reason which the Security Council deemed legitimate, the Security Council might be prompted to assist the party by carrying out positive security guarantees or initiating peacekeeping measures on its behalf. Alternatively, if the reason of the party contemplating withdrawal was deemed illegitimate, the Security Council might then proceed to sanction the party in question for endangering the peace[51c].

The NNWS, in dealing with issues of review, duration and withdrawal, attempted to make more stringent the conditions under which an NPT would remain in force. Their proposals were designed to reduce the durability and increase the conditional nature of an NPT, thus giving themselves additional opportunities to monitor progress towards a more comprehensive arms limitation and security régime and to protect their own interests should progress not occur. This position was entirely consistent with the requirements of the Low Posture Doctrine.

The NNWS which took positions on the issue saw the Review Conference as explicitly related to the obligations of the NWS to continue the process of arms limitation and disarmament, and sought to emphasize this linkage. As noted above, Romania stated this relationship most explicitly by including in its proposed amendment obligating the NWS to adopt specific measures of arms limitation and disarmament, the requirement that "If five years after the entry into force of this Treaty such measures have not been adopted, the Parties shall consider the situation created and decide on the measures to be taken"[51d, 149]. Brazil also offered an amendment to Article V specifying the link between a Review Conference and, *inter alia*, the obligations of the NWS to negotiate arms limitation and disarmament measures[151]. Thus the NNWS expected the Review Conference both to verify the degree to which the NWS were fulfilling their obligations to continue the process of arms limitation and disarmament and to prod the NWS into doing so. As Romanian Ambassador Ecobesco described it:

That provision is intended to be a permanent reminder to the nuclear Powers that they have assumed, in regard to the non-nuclear States and to the entire international community, a commitment which they cannot evade, a commitment which they are bound to respect. Five years after the entry into force of the treaty — a period long enough for the nuclear Powers to show in practice their devotion to the cause of disarmament and peace — the parties to the treaty, in considering the situation created, will be able to draw the appropriate conclusions and to take the required measures[141, 142, 144].

In addition, Romania proposed an amendment to Article V stating that after the first Review Conference: "Such conferences shall be convened thereafter periodically every five years, to review the manner in which the obligations assumed by all Parties to this Treaty are carried out"[149]. Such automatic, mandatory, periodic review conferences would, in Romania's view, provide for "the progressive implementation of the treaty" and facilitate the normal operation of the "verification machinery"[245]. Burma endorsed this Romanian proposal[141].

The NNWS which took positions on the duration of an NPT diverged on the question of whether it should be of indefinite duration. Canada argued that it should be indefinite, since "a treaty of fixed duration would be subject to disintegration at the end of the period" and thus be ineffectual[224]. Switzerland, in a memorandum to the ENDC, expressed the contrary view that an NPT should last for some definite period at the end of which a review conference should decide upon renewal[227]. Italy proposed an amendment representing an intermediate position between unlimited and fixed duration: "This treaty shall have a duration of X years and shall be renewed automatically for any party which shall not have given, six months before the date of expiry of the treaty, notice of its intention to cease to be a party to the treaty"[255]. Both Switzerland and Italy, in contrast to Canada, argued that an NPT of limited duration would, like the Fanfani Proposal of 1965, constitute an incentive to the NWS to take effective measures of nuclear disarmament. They argued that the NNWS could not be expected to forgo their nuclear weapon option indefinitely in the absence of such measures[158, 227].

Finally, in connection with the related question of withdrawal by individual parties, Nigeria proposed an amendment to Article VII whereby the treaty would continue to be of unlimited duration, but each party would have more specific grounds for withdrawal than in the 1967 draft treaties. Such grounds for any party were to be as follows:

(*a*) that the aims of the Treaty are being frustrated;
(*b*) that the failure by a State or group of States to adhere to the Treaty jeopardize the existing or potential balance of power in its area, thereby threatening its security;
(*c*) any other extraordinary events, related to the subject matter of this Treaty, have jeopardized the supreme interests of its country[229].

Thus, non-fulfilment of obligations by parties and non-adherence by threatening non-parties would constitute grounds for withdrawal[226]. Burma endorsed this position with specific reference to the non-fulfilment of obligations by NWS parties with respect to nuclear disarmament[141]. Such broadening of the withdrawal grounds would thus explicitly compensate the NNWS for the unlimited duration of the treaty.

The 1968 identical draft treaties

At the outset of the third stage of the negotiations, the USA and the USSR retained in the January 1968 draft treaties the single review conference and the withdrawal clause provided in the 1967 draft treaties. They did, however, limit the duration of the treaty, which was formerly unlimited, to an initial period of 25 years, after which time "a Conference shall be convened to decide whether the Treaty shall continue in force indefinitely, or shall be extended for an additional fixed period or periods. This decision shall be taken by a majority of the Parties to the Treaty."

On 22 February 1968, the UK resubmitted its previous amendment expanding the scope of the Review Conference so as to include in its purview the purposes of the preamble and, specifically, the preambular language relating to arms limitation and disarmament[51e, 136, 167, 256]. UK Ambassador Mulley justified this amendment on the grounds that "the preamble is still wider than the new Article VI in the disarmament field and indicates in some detail what needs to be done as well as containing an important declaration of intent to achieve at the earliest possible date the cessation of the nuclear arms race"[167]. The amendment was supported by Canada, Ethiopia, FR Germany, Italy, Mexico, Nigeria, Sweden and the UAR, and was subsequently incorporated in the 11 March 1968 joint draft treaty and in the final NPT[155, 173-177, 232].

With respect to the issue of duration, the NNWS, with the exception of Italy and Spain, accepted the provision of the January 1968 draft treaty[176, 257]. This provision was subsequently incorporated in the 11 March 1968 draft treaty and in the final NPT.

With respect to the withdrawal clause, again, most NNWS accepted the provisions of the January 1968 draft treaty. Brazil and Nigeria each proposed amendments providing broader and more hypothetical grounds for withdrawal. Brazil proposed that: "Each Party shall in exercising its national sovereignty, have the right to withdraw from the Treaty if it decides that there have arisen or *may arise* circumstances related with the subject matter of this Treaty which may affect the supreme interests of its country"[258]. Nigeria revised its earlier amendment to read:

> Each Party shall in exercising its national sovereignty have the right to withdraw from the Treaty if it decides that extraordinary events, or *important international developments*, related to the subject matter of this Treaty, *have jeopardized or are likely to jeopardize the national interests of its country*[233].

These attempts to broaden the pretexts for withdrawal were, however, rejected by the NWS and the existing withdrawal provision was subsequently incorporated in the 11 March 1968 draft treaty and in the final NPT.

As in the previous stage of negotiation, however, the scope and frequency of the Review Conference still remained a major issue for the NNWS. Thus, in addition to supporting the British amendment specifically broadening the mandate of the Review Conference, several NNWS proposed periodic Review Conferences in an NPT. The more stringent formulation of this proposal, offered by Italy and Romania and endorsed by FR Germany and Spain, called for automatic, mandatory Review Conferences every five years[164, 176, 232, 257, 259]. On 8 February 1968, Sweden suggested a more moderate amendment to Article VIII: "At the interval of five years thereafter, a majority of the Parties to the Treaty may obtain, by submitting a proposal to this effect to the Depository Governments, the convening of further conferences with the same objective of reviewing the operation of the Treaty"[171]. This formulation was endorsed by Canada, Ethiopia, Mexico, Nigeria and the UK and was subsequently incorporated in the 11 March 1968 joint draft treaty and in the final NPT[136, 173, 175, 177].

The NNWS throughout this stage of the negotiations consistently argued that the conclusions of the Review Conferences and, ultimately, the viability and duration of an NPT, would turn on the question of whether the NWS fulfilled their counterdiscriminatory obligation to negotiate effective measures of arms limitation and disarmament. On 28 February 1968 Nigeria submitted an amendment to Article VI stipulating that "the findings of the review conferences shall be adopted by a majority of the signatory states present"[233, 260]. Thus a majority of the parties might determine whether the purposes and provisions of an NPT were being realized and make this determination "available to those Powers which alone can make the principal aim of the treaty — nuclear disarmament — realizable"[177]. This amendment was not supported by other states. However, it raised a basic quandary which was well summarized by Swedish Ambassador Myrdal:

> An open question remains: namely, what action is suggested to follow if the verdict of a review turns out to be 'unsatisfactory'? It would seem reasonable that, if it is manifest at a review conference that the intentions of the treaty to achieve cessation of the nuclear arms race and to obtain nuclear disarmament have in reality been blatantly disregarded, parties to the treaty may come to regard this as an extraordinary event jeopardizing their own supreme interests . . . [155]

This view explicitly asserts the premises of the Low Posture Doctrine.

The March and May 1968 joint draft treaties

In the March 1968 draft treaty, the USA and the USSR incorporated the two earlier amendments of Sweden and the UK providing for periodic Review Conferences which, *inter alia*, would be specifically authorized to review the purposes of the preamble relating to arms limitation and disarmament. The new Article VII, paragraph 3 provided that:

> Five years after the entry into force of this Treaty, a conference of Parties to the Treaty shall be held in Geneva, Switzerland, in order to review the operation of this Treaty with a view to assuring that the purposes of the Preamble and the provisions of the Treaty are being realized. At intervals of five years thereafter, a majority of the Parties to the Treaty may obtain, by submitting a proposal to this effect to the Depositary Governments, the convening of further conferences with the same objective of reviewing the operation of the Treaty.

In presenting these amendments, the NWS argued that the provisions for periodic review of the NPT gave 'further force' to the obligations undertaken under the preamble and Article VI concerning arms limitation and disarmament, thereby again conceding the logic of the Low Posture Doctrine[168, 183].

Most NNWS, in the final stages of the negotiations on an NPT, did not press for further changes in these provisions for review, duration and withdrawal[7e]. They had already succeeded in introducing more flexibility into these provisions than had existed in the earlier draft treaties. In particular, the NNWS had been instrumental in clearly stipulating that the operation of an NPT, and specifically the counterdiscriminatory obligations of the NWS in Articles IV, V and VI, would be subjected to careful, periodic review[7f, 182, 186]. Thus, no further changes were made in the provisions relating to review, duration and withdrawal and they were incorporated into the final NPT.

IV. Conclusion

This historical review of the negotiations on the NPT has described how differently the NWS and the NNWS define the arms limitation and security régime which should buttress an NPT. Despite some slight convergence as a result of the negotiations, the obligations, expectations, risks and privileges ultimately imposed upon these two classes of states in the final NPT remained very different both in their substance and in the way in which they were legally binding. In the view of many NNWS, this different allocation fell short of the minimum "acceptable balance of mutual responsibilities and obligations" required of an NPT by Resolution 2028 (XX)[143, 176, 189, 240].[2]

The first three articles of the NPT constituted the major proposals articulated by the NWS in their various draft treaties. Article I requires the NWS only to forgo immediately their right to disseminate nuclear explosive devices or control over them to NNWS. This obligation is not subject to control and inspection. Thus under the treaty, the NWS can continue to enjoy the prestige and security stemming from their possession of nuclear weapons unconstrained by any limits upon their rights to develop, test, produce, deploy and utilize them.

On the other hand, Article II requires the NNWS immediately to forgo manufacturing or otherwise acquiring nuclear explosive devices even out of their nationally owned fissile material, resources and facilities. Article III requires that control and inspection agreements be negotiated with the

[2]Both the UAR and Romania proposed that Resolution 2028 (XX) be included in the treaty.

International Atomic Energy Agency on all peaceful nuclear activities of the NNWS, theoretically within two years after the entry into force of the NPT or within 18 months after ratification, for NNWS ratifying the NPT after its entry into force. Thus, these first three Articles collectively impose stringent specific, physical and immediate or near-term obligations upon the NNWS. By so doing, they treat the two classes of states so disproportionately as to constitute a clearly discriminatory allocation of obligations.

The subsequent four articles of the NPT and Security Council Resolution 255 constitute attempts by the NNWS to introduce into the NPT some counterdiscriminatory obligations incumbent upon the NWS. Articles IV and V, which are not addressed here, are designed to provide NNWS with guaranteed access to international cooperation in civilian nuclear development and peaceful nuclear explosions. Article VI requires the NWS to pursue further negotiations on arms limitation and disarmament 'in good faith'. Article VII declares support for the principle of denuclearized zones. Security Council Resolution 255 and its associated unilateral declarations provide joint positive security guarantees through the Security Council and unilateral or several positive security guarantees through Article 51 of the UN Charter.

All these obligations incumbent upon the NWS are, however, hortatory, putative, contingent and subject to fulfilment only at some unspecified future point in time. It is precisely the nature of such hortatory obligations which exhorts certain actors to engage in certain discretionary behaviour in the future, that they are at once legally binding and unenforceable[261]. In particular, Article VI imposes no specific, physical and immediate or near-term requirements upon the NWS to adopt any particular arms limitation and disarmament measures constraining their rights to develop, test, produce, deploy and utilize nuclear weapons. Article VII does not commit the NWS to undertake specific negative security guarantees with respect either to the existing denuclearized zone in Latin America or to future denuclearized zones which might subsequently be established in other regions. Although no situation requiring the Security Council to consider measures provided for in Resolution 255 has yet occurred, thereby leaving its scope and effectiveness untested, many NNWS have suspected that Security Council Resolution 255 would not provide effective positive security guarantees[73a, 92]. Furthermore, with the exception of the declaratory support in Article VII for the principle of denuclearized zones, the NPT and Security Council Resolution 255 are in no way responsive to the continuing demands of many NNWS for additional negative security guarantees forswearing certain uses or threatened uses of nuclear weapons by NWS parties to an NPT. Thus, with respect to the comprehensive arms limitation and security régime of interest to the NNWS, these counterdiscriminatory obligations do not substantially rectify the discriminatory allocation of rights and obligations stipulated in the first three articles of the NPT.

It was noted above that the participating NWS granted that the assessment of the NPT as a discriminatory treaty was to some degree valid. However, throughout the negotiations and since their conclusion, the participating NWS have defended the differential allocation of rights and obligations in the NPT in three ways. First, they have argued that a partial, limited NPT which merely constrains the proliferation of nuclear weapons to additional nation-states provides an absolute gain for the entire international political system. It does not, therefore, require compensatory and balancing concessions from the NWS in the form of a *quid pro quo* within the treaty itself as if it were a commercial contract[164, 262].

Second, given their commitment to the absolute value of an NPT, the NWS opposed for tactical reasons a treaty encompassing the more comprehensive range of arms limitation and security issues of interest to the NNWS, lest incorporating such contentious issues should delay and perhaps make impossible any agreement[136, 263]. Although periodically acknowledging that the absence of this more comprehensive range of measures constituted an arguably imbalanced allocation of rights and obligations, the NWS have nevertheless defended it as tactically necessary in the short run. To quote British Ambassador Chalfont on this point: "'Do as I say, not as I do' is a logically indefensible precept, but in an imperfect world it may be a necessary one for a particular short period of history"[105].

Third, the NWS have argued that a partial, limited NPT would and should facilitate subsequent agreements on a more comprehensive range of arms limitation, disarmament and security measures[131, 262]. The NWS have repeatedly admitted that such agreements would have to be reached if the NPT were to remain viable. Indeed, the NWS have suggested that the NNWS would be instrumental in bringing pressure to bear in the NPT Review Conferences towards this end. To quote US Ambassador Goldberg speaking in 1968 on this point:

> I turn to the fourth question: will this treaty help bring nearer an end to the nuclear arms race, and actual nuclear disarmament, by the nuclear-weapon States, and will it help achieve general disarmament?
>
> Again the answer is yes. Once again, it was chiefly at the initiative of the non-nuclear States that this problem was directly addressed in the operative section of the treaty by the insertion of Article VI. In that article all parties "undertake to pursue negotiations in good faith" on these further measures. This is an obligation which, obviously, falls most directly on the nuclear-weapon States.
>
> Ideally, in a more nearly perfect world, we might have tried to include in this treaty even stronger provisions — even perhaps an actual agreed programme — for ending the nuclear arms race and for nuclear disarmament. But it was generally realized in the Eighteen-Nation Disarmament Committee that, if we were to attempt to achieve agreement on all aspects of disarmament at this time, the negotiating difficulties would be insurmountable and we should end by achieving nothing.
>
> However, this treaty text contains, in article VI, the strongest and most meaningful undertaking that could be agreed upon. Moreover, the language of this article indicates a practical order of priorities — which was seconded in the statement read on behalf of the Secretary-General — headed by "cessation of the nuclear arms race at an early date" and proceeding next to "nuclear disarmament" and finally to

''general and complete disarmament under strict and effective international control'' as the ultimate goal.

Let me point out that further force is imparted to article VI by the provision in article VIII for periodic review of the treaty at intervals of five years, to determine whether the purposes of the preamble and the provisions of the treaty are being realized. My country believes that the permanent viability of this treaty will depend in large measure on our success in the further negotiations contemplated in Article VI[264].

In sum, while asserting the limited, partial and inherently discriminatory nature of the NPT, the NWS have repeatedly acknowledged their counterdiscriminatory obligations to achieve, over time, the more comprehensive arms limitation and security régime of interest to the NNWS. The NNWS, for their part, have continued to urge that the NWS begin as soon as possible to buttress and amplify the NPT with a more comprehensive arms limitation and security régime in order to redress the discrimination inherent in the NPT, ensure its viability over time, and minimize the proliferation of nuclear weapons. These contrasting positions have been repeatedly expressed in the ENDC/CCD and at the UN. They were perhaps most clearly restated at the Review Conference of the Parties to the NPT held in Geneva in May 1975.

4. The 1975 Review Conference of the Parties to the NPT

The first Review Conference of the NPT was held, pursuant to Article VIII, "in order to review the operation of this Treaty with a view to assuring that the purposes of the Preamble and the provisions of the Treaty are being realized"[265, 266]. Fifty-eight parties to the NPT attended, as did seven states which had signed but not yet ratified it. The Conference elected Inga Thorsson, Swedish Under-Secretary of State, as its President, and established two main committees: Committee I for review of the implementation of the NPT as it relates *inter alia* to the question of arms limitation, disarmament and security guarantees; and Committee II for review of the implementation of the NPT as it relates to the question of safeguards, peaceful uses of nuclear energy and peaceful nuclear explosions.

In debating the relationship between the arms limitation and security policies of the NWS and the future acquisition of nuclear weapons by the NNWS, it was the NNWS, rather than the NWS, which submitted substantive proposals at the Review Conference.[1] These initiatives were consistent with the declared position of many NNWS that the NPT needed to be amplified in order to constitute a comprehensive and equitable arms limitation and security régime. The NWS, for their part, stressed that the basic purposes of the NPT had been fulfilled in that the undertakings in Articles I and II not to disseminate or acquire nuclear explosive devices had been observed by all parties[269]. The NWS further defended their existing arms limitation and security policies as consistent with their obligations under the treaty, promised future efforts according to their own agenda and at their own pace, but resisted all substantive initiatives of the NNWS in these areas. The various proposals made to the Review Conference are summarized below, again contrasting the positions of the NWS and NNWS

[1]The Review Conference could have adopted at least three types of final document as a result of its work: a general declaration; various resolutions relating to substantive items on its agenda; and various additional protocols which the Conference might "recommend for signature and ratification in order to supplement the provisions of the Treaty with some urgent measures aimed at promoting a wider acceptance of the Treaty"[267]. After a 26-day session, the Conference failed to reach recommendations acceptable to both the NWS and the NNWS. Thereupon, a compromise draft Final Declaration was prepared by Conference President Thorsson. This compromise was adopted by consensus and was amplified by interpretive statements annexed to it[268]. The Review Conference was unable to agree on any further substantive resolutions implementing the NPT. At the insistence of their various sponsors, however, certain draft resolutions annexing originally submitted draft additional protocols and other draft resolutions were also annexed to the Final Declaration.

with respect to arms limitation and disarmament measures, security guarantees and further provisions of review.

I. Arms limitation and disarmament measures

At the Review Conference, as during the negotiation of the NPT, the NWS conceded the underlying logic of the Low Posture Doctrine that "the long-term success of non-proliferation depended to an important degree on the implementation of Article VI, as well as on other important factors"[270]. From the outset of the Review Conference, however, the USA, the USSR and their respective close military allies refused to consider any proposals which imposed additional obligations on the NWS to pursue negotiations on arms limitation and disarmament measures. The NWS justified this position on three grounds. First, they stressed that the progress made in various arms limitation and disarmament negotiations since the entry into force of the NPT in 1970 constituted the specific implementation of their obligations under Article VI. As evidence of accomplishment, both the USA and the USSR listed in their opening speeches various bilateral and multilateral agreements which had been struck since 1970, including the 1971 Sea-Bed Treaty, the 1971 Biological Weapons Convention, the 1971 Agreement on the prevention of incidents on and over the high seas, the 1971 Agreement on measures to reduce the risk of outbreak of nuclear war between the USA and the USSR, the 1972 ABM Treaty, the 1972 Interim Agreement between the USA and the USSR on certain measures with respect to the limitation of strategic offensive arms, the 1973 US-Soviet Agreement on the prevention of nuclear war, the 1973 Protocol to the Agreement on the prevention of incidents on and over the high seas, the 1974 Protocol to the ABM Treaty, the 1974 Treaty on the limitation of underground nuclear weapons tests, and the 1974 so-called Vladivostok Accords on the limitation of strategic offensive arms[269, 271-273]. The USA and the USSR also cited continuing negotiations on strategic arms limitations, the control of chemical weapons, the limitation of peaceful nuclear explosions in connection with the Treaty on the limitation of underground nuclear weapon tests, and the reduction of military forces and armaments in Central Europe. The Soviet Union and its allies also took the occasion to press for certain initiatives which they had supported at the United Nations during this period, including: a World Disarmament Conference, Resolution 2936 (XXVII) on the non-use of force in international relations and prohibition of use of nuclear weapons, the reduction of military budgets of states which are permanent members of the Security Council and the transfer of part of these savings to underdeveloped countries, and the prohibition of environmental warfare.

In summarizing this record, Ambassador Ikle of the USA concluded his opening remarks with the claim that:

> In the five years that had elapsed since the Treaty on the Non-Proliferation of Nuclear Weapons had come into effect, far more had been accomplished in the control of nuclear arms than in the preceding 25 years. The Treaty had proved to be both a prerequisite and a catalyst for progress towards nuclear disarmament. The disarmament process was under way and it was up to all States to encourage and sustain it[269, 274].

Furthermore, the USA and the USSR stressed their commitment to détente and to continued progress in the arms limitation and disarmament negotiations, particularly on SALT II and on a CTB. Thus, given their record to date and their commitment to further progress, the two powers claimed that their obligations under Article VI and the preambular paragraphs relating to arms limitation and disarmament were being fulfilled[275].

Second, the NWS also argued that all parties, not only the NWS parties, were responsible for the achievement of both nuclear and non-nuclear arms limitation and disarmament measures under the NPT. They thus urged other states to fulfil their own commitments[272, 276]. The Soviet Union in particular stressed the need for a universal CTB which would also require ratification by China and France[273].

Third, the NWS consistently focused on the need to gain additional — ideally universal — adherence to the NPT. They argued, as they had during negotiation of the NPT, that significant arms limitations would not be possible in a world of many NWS. Therefore, progress between the NWS in arms limitation and disarmament negotiations would depend critically upon the control of horizontal proliferation[275, 277]. The central concern of the Review Conference should, in the view of the NWS and their close military allies, be to induce additional NNWS to adhere to the NPT, rather than to pressure the NWS parties to pursue arms limitation and disarmament agreements. In this regard, the NWS were consistent with the position they first elaborated in 1965: that the NPT was a limited, collateral measure explicitly designed to limit the horizontal proliferation of nuclear weapons.

Most of the NNWS at the Review Conference simply rejected the basic proposition of the NWS that they were fulfilling their obligations under Article VI. Rather, they argued that the agreements and negotiations cited by the NWS did not constitute real progress towards effective measures of arms limitation and disarmament as required by Article VI. While some NNWS welcomed SALT I, the Vladivostok Accords and the Threshold Test Ban Treaty as setting ceilings on the nuclear forces of the USA and the USSR, most argued that these limits merely ratified pre-existing military force levels. Thus, they permitted substantial quantitative increases and qualitative improvements in the two powers' nuclear inventories and 'institutionalized' the arms race. The NNWS argued further that agreements relating to environments of peripheral military concern, such as the sea-bed, and various agreements managing the deployment of military forces did not really constitute effective measures of arms limitation as required under Article VI[276, 278-284]. To quote Ambassador Roberts of New Zealand on this point:

It was small wonder that the countries outside the Treaty remained unconvinced that the nuclear-weapon parties were serious in their intention to give effect to their undertakings. The most valid test of progress was surely to ask whether or not there were fewer nuclear weapons in existence today than there had been in 1970; whether or not there had been any significant abatement in nuclear weapons testing during that period; and whether or not there had been any halt in the further refinement and sophistication of nuclear weapons. The answer to all three questions was patently no. The limited and peripheral agreements negotiated so far gave little ground for reassurance[283].

Having rejected the performance of the NWS to date, many NNWS moved to press the NWS to reach agreements on effective measures of nuclear arms limitation and disarmament. They defended this action on the grounds that NNWS, particularly the non-aligned, had always played such a catalytic role in the field of arms control and disarmament[273].

The NNWS reiterated three arguments which they had first articulated during negotiations on the NPT. First, the NWS in particular had a binding legal obligation to pursue effective measures relating to the cessation of the nuclear arms race and to nuclear disarmament at an early date. Second, the NWS were obligated to agree upon specific measures within some reasonable time period. Third, such agreements were necessary if horizontal proliferation were to be constrained.

With respect to the binding nature of the legal obligation, the NNWS explicitly rejected the argument of the NWS that Articles I and II constituted the primary continuing purposes of the NPT. Instead, they contended that nuclear arms limitation and disarmament were at least of comparable if not greater importance[269, 276, 278, 280, 281, 285]. In sum, the control of horizontal and vertical proliferation were the dual and inextricable objectives of the NPT.

Furthermore, the NNWS argued that this obligation to pursue effective measures of nuclear arms limitation and disarmament fell primarily upon the NWS parties and that the USA and the USSR were particularly instrumental. Given their lack of progress, the USA and the USSR were thus obligated to "fill the gaps" in the NPT, thereby creating "the acceptable balance of mutual obligations and responsibilities of the nuclear and non-nuclear Powers" required by Resolution 2028(XX). To quote Ambassador Fartash of Iran on this point:

It would be useless to prevent horizontal proliferation of nuclear weapons capabilities if vertical proliferation of the stockpiles of the nuclear Powers was allowed to continue unhampered. Article VI of the Treaty met that concern; the non-nuclear-weapon States Parties to the Treaty had accepted the division of the world into nuclear-weapon and non-nuclear weapon States only on condition that the nuclear Powers committed themselves to effective nuclear disarmament measures [279].

Second, the NNWS argued that the NWS were obligated not only to pursue, but ultimately to agree to, some specific, tangible measures of nuclear arms limitation and disarmament within a reasonable time period. The achievement of a Comprehensive Test Ban Treaty was repeatedly identified as the single most compelling obligation of the NWS, mandated as it had been since the Partial Test Ban Treaty of 1963. The gradual reduction of nuclear weapon capabilities below the ceilings reached in the Vladivostok Accords was also stressed. Other specific measures which were propounded included limits on missile flight tests and the cut-off of production of fissile materials for nuclear weapons.

With respect to the period in which the NWS were obligated to achieve such agreements, a major discrepancy between the obligations of the two classes of states parties to the NPT lies in the timing of their obligations: the NNWS undertake to fulfil their obligations upon adhering to the treaty while the NWS undertake to do so only at some unspecified 'early date' in the future. At the Review Conference, it was widely argued by the NNWS that the early date had arrived and that implementation of such measures of nuclear arms limitation and disarmament was either urgent or already overdue[273, 276, 279-286].

Lastly, the NNWS argued that, consistent with the Low Posture Doctrine, effective measures of arms limitation and disarmament were a necessary, if not sufficient, condition of non-proliferation of nuclear weapons. Not only did lack of progress towards these goals provide NNWS not yet parties to the NPT with a convenient excuse for non-adherence, but such unwillingness on the part of the NWS gradually to reduce their nuclear weapons also made it unlikely that NNWS would continue to forswear them indefinitely[276, 279, 281-283].

At the end of the general debate, Conference President Thorsson summed up the NNWS position as follows:

> While everyone has agreed upon the success of the NPT, in that no non-nuclear weapon State adhering to the Treaty has got possession of nuclear weapons, the general view among a majority of the non-nuclear-weapon States, as emerging from their statements is very clearly that the nuclear-weapon States have not achieved results to the satisfaction of the non-nuclear weapon States parties in efforts towards genuine nuclear disarmament. It seems to me that an enlightened world opinion, reflected in this case, in statements by non-nuclear-weapon States, rather impatiently awaits concrete and binding results of on-going bilateral negotiations, aiming at ending the quantitative and qualitative arms race, and reducing substantially the levels of nuclear armaments. Many have referred to the need for a time-table for results to be achieved through these negotiations. The agreement on a comprehensive test ban is clearly recognized as a most decisive element in these efforts. A least common denominator is apparent in the statements: Article VI must be implemented, in letter and in spirit[287].

Various NNWS adopted two different approaches to implement their position. One, initiated by Mexico, was the drafting of Additional Protocols I and II to the NPT which addressed both the contention of the

NWS that nuclear arms limitation and disarmament could not proceed unless horizontal proliferation were constrained, and the contention of the NNWS that such proliferation could not be constrained without effective measures of nuclear arms limitation and disarmament. Thus, both Additional Protocols linked these two inextricable objectives by automatically relating specific and increasingly stringent measures of nuclear arms limitation to be undertaken by NWS parties to the NPT to adherence by increasing numbers of NNWS to the NPT.

Additional Protocol I, sponsored by 20 NNWS — Bolivia, Ecuador, Ghana, Honduras, Jamaica, Lebanon, Liberia, Mexico, Morocco, Nepal, Nicaragua, Nigeria, Peru, the Philippines, Romania, Senegal, the Sudan, Syria, Yugoslavia and Zaire — was designed to achieve the complete cessation of all nuclear weapon tests. The USA and the USSR would undertake

> to decree the suspension of all their underground nuclear tests for a period of ten years, as soon as the number of Parties to the Treaty reaches one hundred; . . . to extend by three years the moratorium . . . each time that five additional states become party to the Treaty; . . . [and] to transform the moratorium into a permanent cessation of all nuclear weapon tests, through the conclusion of a multilateral treaty for that purpose, as soon as the other nuclear weapon States indicate their willingness to become parties to said treaty.

The protocol would be subject to ratification by the three depositary states of the NPT and enter into force when ratified by two of them[288].

Additional Protocol II, sponsored by all the above NNWS except the Philippines, was designed to achieve substantial reductions in the nuclear weapon capabilities of the USA and the USSR. Having agreed to certain ceilings on their respective nuclear capabilities in the Vladivostok Accords, the USA and the USSR would

> undertake, as soon as the number of Parties to the Treaty has reached one hundred
> (*a*) To reduce by fifty per cent the ceiling of 2,400 nuclear strategic delivery vehicles contemplated for each side under the Vladivostok accords;
> (*b*) To reduce likewise by fifty per cent the ceiling of 1,320 strategic ballistic missiles which, under these accords, each side may equip with multiple independently targetable warheads (MIRV's); . . . [and to] also undertake, once such reductions have been carried out, to reduce by ten per cent the ceilings of 1,200 strategic nuclear delivery vehicles and of 660 strategic ballistic missiles that may be equipped with multiple independently targetable warheads (MIRV's), each time that ten additional States become Parties to the Treaty.

The Protocol would enter into force when both the USA and the USSR had ratified it[289].

Each Protocol was justified on the grounds that it would not jeopardize the security of the affected NWS. In the case of the cessation of nuclear weapon tests, it would still leave the USA and the USSR with 'indisputable superiority' in nuclear weapon technology over all minor NWS and NNWS. With respect to the reduction of strategic delivery vehicles, it would not affect the balance between the two powers. Moreover, even if the NPT gained universal adherence, the USA and the USSR would retain 600 strategic nuclear delivery vehicles, 330 of which might be equipped with MIRVs. Such residual inventories would be substantially in excess of the combined capabilities of all minor NWS. In addition, each NWS would still have the right of withdrawal should the supreme interests of its country require it[283, 290, 291].

A second approach to achieving nuclear arms limitation and disarmament measures was proposed by Sweden in a working paper supported by several NNWS, including Australia and Iran[276, 279, 282, 292, 293]. In the words of Swedish Ambassador Hamilton:

> The Conference should first agree on the immediate goals to be achieved in the field of nuclear disarmament; second, a time-table should be established for achieving those aims on the basis of a realistic proposal by the two nuclear States concerned; third, a second review conference should be convened for 1980, which would be a factor in strengthening the non-proliferation regime. It should consider what effect had been given to the provisions of Article VI, in respect of negotiations on the reduction of nuclear arms, what reductions had actually been achieved, and the situation with regard to an underground test ban[285].

Sweden stated that its approach had the same general aims as the two additional protocols without their automatic linkage to additional adherence to the NPT.

Sweden proposed a precise timetable whereby the NWS would begin immediate negotiations to reach a comprehensive test ban before the second Review Conference in 1980. As an interim measure, they were to place a moratorium on all underground nuclear tests for a specified period. In addition, the USA and the USSR were to conclude a SALT II agreement based upon the Vladivostok Accords by the end of 1975. Thereafter, they were to begin negotiations at SALT III on reducing their strategic nuclear forces. Such negotiations were to result in an agreement before a second Review Conference in 1980.

In an additional draft resolution relating to arms limitation and disarmament, Romania urged the NWS parties to expedite negotiations and agreements at the CCD and recommended the establishment of a system of retrieval, distribution, assessment and analysis of information on armaments and disarmament issues within the United Nations[294].

Most of the subsequent attention to this subject at the Review Conference focused upon the two draft Additional Protocols. The NWS and their close military allies rejected them out of hand, on three grounds. First, the NWS asserted that the protocols, as well as the sort of precise timetable envisaged in the Swedish working paper, were beyond the terms of reference of the Review Conference. In particular, the protocols were procedurally flawed since they sought in effect to amend the NPT without conforming to the stipulated procedures for amending the treaty as elaborated in Article VIII[274]. Moreover, such proposals intruded upon matters of concern only to the NWS. The USSR flatly asserted that the protocols were unacceptable and inadmissible interferences in US-Soviet relations, imposing unwarranted restrictions on only two parties to the NPT. Many allied states concurred, noting that it was not normal for third countries to give detailed instructions to states concerning the manner in which their negotiations should be conducted[270, 272, 273, 275].

Second, the NWS and their allies argued that both a CTB and SALT raised technically complex and serious issues which could not be resolved according to "arbitrary timetables", "random figures", "mathematical formulae", and "artificial deadlines"[272, 274, 295]. Neither the Swedish timetable nor Additional Protocol I addressed difficult and hitherto divisive questions of universal adherence to a CTB, verification, and the treatment of peaceful nuclear explosives. And Additional Protocol II based its procedures upon the Vladivostok Accords, which were themselves still subject to continuing negotiation. In sum, the protocols were unrealistic and unproductive, and the USA and the USSR each stressed that the SALT and CTB negotiations would continue to be carried out at a pace consistent with their own perceived concepts of national interests and security[270, 293, 296].

Finally, the NWS and their allies argued that linking a comprehensive test ban and strategic arms limitations to additional adherence to the NPT was "arbitrary", and unconvincing "gimmickry". They refuted the Low Posture Doctrine's logic that such measures were necessary to minimize proliferation and in fact questioned whether such measures would attract new signatories to the NPT. They also questioned, conversely, why some magic number of NPT adherents should make agreements on complex issues like a CTB or SALT more viable[270, 272].

Ambassador Garcia Robles of Mexico spoke on behalf of the Group of 77 NNWS at the Review Conference regarding the two draft Additional Protocols, refuting the charges of the NWS that the sponsors sought to amend the treaty illegally or impose new obligations upon the NNWS parties. He asserted that the protocols were merely intended to carry out the existing obligations of the relevant parties to the NPT under Article VI to reach effective measures of nuclear arms limitation and disarmament at an early date. Ambassador Garcia Robles also offered to change Additional Protocol I so as to stipulate that several NNWS near the nuclear threshold be among the additional adherents required before the test moratorium could come into being[272, 273, 275]. However, there was no possibility of serious negotiation with the NWS on the various proposals of the NNWS relating to the further implementation of Article VI, and the texts remained unchanged.

Given the impasse on the issue of further arms limitation and disarmament measures which might be related to the NPT — among the most intractable issues of the Review Conference — Conference President Thorsson incorporated several points in her draft Final Declaration reviewing Article VI. This compromise language, which proved acceptable to the Conference, outlined the concerns of the NNWS while avoiding specific criticisms of the NWS. First, while welcoming the various agreements on arms limitation and disarmament which had been reached over the past few years "as steps contributing to the implementation of Article VI of the Treaty", the Conference urged greater efforts by all parties, but particularly the NWS, "to achieve an early and effective implementation of Article VI". The Conference urged the NWS parties to reach an early agreement on an effective CTB and cited the desire of "a considerable number of delegations" — that is, the NNWS — that the NWS parties

as soon as possible enter into an agreement, open to all states and containing appropriate provisions to ensure its effectiveness, to halt all nuclear weapons tests of adhering states for a specified time, whereupon the terms of such an agreement would be reviewed in the light of the opportunity, at that time, to achieve a universal and permanent cessation of all nuclear weapons tests.

Meanwhile, the Conference urged all NWS signatories to the Threshold Test Ban Treaty to limit the number of their underground tests to a "minimum".

The Conference also appealed to the USA and the USSR to try to conclude at the earliest possible date the new agreement outlined in the Vladivostok Accords and to commence follow-on negotiations on "further limitations of and significant reductions in" their nuclear inventories as soon as possible following such a SALT II agreement. Finally, the CCD was urged to increase its efforts to achieve effective disarmament agreements[268].

Ambassador Garcia Robles stated that Mexico and the other states sponsoring the two Additional Protocols on arms limitation and disarmament measures had joined in the consensus on the Final Declaration on the clear understanding that the texts of the draft resolutions recommending the Additional Protocols would be annexed to the Final Declaration of the Review Conference[297, 298]. They were so annexed: clear testimonies to the inability of the NWS and the NNWS to agree upon how further measures of arms limitation and disarmament should best be related to the NPT.

II. Security guarantees

In discussing security guarantees at the Review Conference, the NWS, supported by their close military allies, remained consistent with the positions they had taken on the NPT and with the logic of the High Posture Doctrine. Thus they refused to entertain any proposals which imposed on them additional obligations or restrictions relating to their deployment and use of nuclear weapons[270, 273, 293, 296].

With respect to positive security guarantees, both the USA and the USSR argued that the existing guarantees incorporated in Security Council

Resolution 255 and its associated unilateral declarations were the maximum feasible to which they could commit themselves without jeopardizing other important security interests. Moreover, while granting that Resolution 255 had never been tested, they argued that it constituted an important political undertaking which by this very token had apparently provided some protection for the NNWS. Thus, they concluded, it deserved to be reaffirmed, rather than amplified or replaced.

With respect to negative security guarantees, the NWS reaffirmed Article VII of the NPT which endorsed the principle of nuclear-free zones. They argued that such zones would enhance the security of NNWS in appropriate regions, but that discussion of specific zones was the responsibility of the regional states in question and thus outside the purview of the Review Conference. Moreover, both the USA and the USSR refused to undertake in advance commitments not to use nuclear weapons in such zones. With respect to universal negative security guarantees, the USA specifically and the USSR implicitly rejected them.

The USA and the USSR differed somewhat on additional approaches for meeting the various security needs of NNWS. The USA and its allies stressed the utility of the positive security guarantees provided through mutual security relationships for protecting certain NNWS allies. The USA, joined by South Korea and FR Germany, stressed that such alliance relationships might well be undermined by various proposals for universal negative security guarantees. This position was consistent with long-standing US opposition to negative security guarantees and no-first-use undertakings.

Moreover, the USA argued that the security of non-aligned NNWS was not significantly threatened by possible nuclear attacks from the NWS parties to the NPT, and that, therefore, negative security guarantees would be of slight utility to them. Rather, such NNWS were more likely to be threatened by the possibility of conventional or future nuclear conflicts with hostile neighbours. Thus, the USA recommended concentrating on more "practical" policy instruments such as encouraging adherence to the NPT by additional NNWS, resolving long-standing regional disputes, and designing conventional arms limitation arrangements. These, rather than negative security guarantees, were appropriate means of meeting the security objectives of non-aligned NNWS[265, 270, 273, 296].

The USSR and its allies, alternatively, stressed more general declaratory policies. They urged that the UN Security Council make legally binding General Assembly Resolution 2936 (XXVII) on "the renunciation of the use or threat of force in international relations and the permanent prohibition of the use of nuclear weapons" which was passed in 1972 at their initiative. This resolution would make the non-use of nuclear weapons contingent upon the non-use of force, thereby permitting the first use of nuclear weapons in the event of conventional war. Thus it did not constitute a no-first-use undertaking which, of course, would incorporate negative security guarantees to NNWS under certain conditions. However, such a legal undertaking as Resolution 2936 (XXVII) would, in the Soviet view, contribute to the security of all states, allied or non-aligned, and whether or not they possessed nuclear weapons. In the same regard, the USSR stressed the importance of the 1973 US-Soviet Agreement on the Prevention of Nuclear War[265, 273, 293, 299].

The NNWS as a class of states face a wide variety of perceived threats to their security. Thus, at the Review Conference, consistent with positions taken during negotiations on the NPT as well as with the logic of the modified Low Posture Doctrine, virtually all of the NNWS advocated one or another of the wide range of positive and negative security guarantees by which NWS guarantors might satisfy their various military security objectives. With the exception of certain close military allies of the NWS, the NNWS viewed the forms of guarantees and alternative security policies proposed by the NWS as providing inadequate protection of NNWS. Moreover, they did not rectify the discriminatory allocation of obligations and rights inherent in the NPT. Thus, the NNWS consistently pressed for stronger forms of positive and negative security guarantees than the NWS were willing to provide.

With respect to positive security guarantees, the NNWS reiterated several objections to Security Council Resolution 255 and its associated unilateral declarations which they had raised in the 1968 Resumed 22nd General Assembly Session on the NPT. First, many NNWS asserted that the guarantees provided therein were redundant and/or incomplete, since they merely reaffirmed existing UN Charter obligations to assist states under attack, irrespective of the weapons used. In addition, Resolution 255 undertook to take action only after a nuclear attack or threat of attack had actually occurred. Thus, it could not prevent such use or theatened use of nuclear weapons. Second, many NNWS doubted the efficacy of Security Council Resolution 255 now that all existing NWS, with the arguable exception of India, were, as of 1971, Permanent Members of the Security Council. Were any NWS to use, or threaten to use, nuclear weapons, it would inevitably veto the exercise of Resolution 255 against itself. Third, in the unlikely event that Security Council Resolution 255 were ever invoked, no consultation with the 'victimized' state would be required before the intervention of NWS on its behalf. This form of hegemonic behaviour on the part of the NWS guarantors was unacceptable to certain non-aligned NNWS, which argued that such intervention should be specifically requested by a NNWS before it could be undertaken by the NWS guarantors[265, 276, 280-284, 286].

Given the flawed nature of existing positive security guarantees, some NNWS simply admitted that, outside of military alliances, no stronger forms were possible in the current political circumstances. Other NNWS, however — such as Egypt, Ghana, the Philippines, Syria and Yugoslavia — urged stronger and more legally binding positive security guarantees from the NWS in order to meet the pressing needs of certain NNWS for protection against hostile states already or potentially armed with nuclear weapons[281, 283, 284, 286].

In addition to positive security guarantees, many NNWS argued, just as they had during negotiations on the NPT and consistent with the Low Posture Doctrine, that the NWS should delimit deployments and permissible uses of nuclear weapons. Given the continued existence of their vast nuclear weapon capabilities, the NWS should at least limit the degree to which they utilize these weapons in the conduct of foreign policy. Such constraints would be particularly apt in the form of negative security guarantees insofar as they affected the interests of NNWS parties to the NPT, since such guarantees would explicitly contribute to "the acceptable balance of mutual obligations and responsibilities of the nuclear and non-nuclear Powers" required by Resolution 2028 (XX). The NNWS also argued that all such deployment and use constraints would attract additional NNWS to adhere to the NPT.

At the end of the general debate, Conference President Thorsson summed up the NNWS positions on security guarantees as follows:

I have interpreted the numerous references to the matter of the security of non-nuclear-weapon States in a world still armed with nuclear weapons as a recognition of the inadequacy of Security Council resolution 255 (1968). There seems to be a close to unanimous feeling among non-nuclear-weapon States that a way should be found for the nuclear-weapon States to pledge, as a legally binding commitment, the non-use of nuclear weapons or threat of nuclear weapons against a non-nuclear-weapon State party to the Treaty. Some delegations have linked such a pledge to the establishing of nuclear-free zones, but I have felt a rather general recognition of the need for increased safety measures against nuclear attacks, or threats of nuclear attacks, on the part of those non-nuclear-weapon States which have through adherence to the Treaty foreclosed their nuclear options[287].

To implement their positions, various NNWS proposed three different approaches. The first was a broad-gauged proposal on security assurances, draft Additional Protocol III, which was initiated by Romania and co-sponsored by Bolivia, Ecuador, Ghana, Mexico, Nigeria, Peru, Senegal, the Sudan, Yugoslavia and Zaire. It sought to introduce as a legal obligation within the framework of the NPT itself a composite undertaking by the NWS parties to provide a universal negative security guarantee, with conditional application to NNWS with nuclear weapons on their territories; to respect future nuclear-free zones; and to provide more binding positive security guarantees. Thus, Additional Protocol III stated that:

Article 1. [The nuclear weapon States Parties to the NPT] solemnly undertake

(*a*) never and under no circumstances to use or threaten to use nuclear weapons against non-nuclear-weapon States Parties to the Treaty whose territories are completely free from nuclear weapons, and,

(*b*) To refrain from first use of nuclear weapons against any other non-nuclear-weapon States Parties to the Treaty.

Article 2. They undertake to encourage negotiations initiated by any group of States parties to the Treaty or others to establish nuclear weapon free zones in their respective territories or regions, and to respect the statute of nuclear weapon free zones established.

Article 3. In the event a non-nuclear-weapon State Party to the Treaty becomes a victim of an attack with nuclear weapons or of a threat with the use of such weapons, the States Parties to this Protocol, at the request of the victim of such threat or attack, undertake to provide to it immediate assistance without prejudice to their obligations under the United Nations Charter.

The Protocol would be subject to ratification by the three depositary states of the NPT and enter into force when ratified by two of them. The duration of the protocol and provisions for withdrawal would be the same as for the NPT itself[276, 279-284, 286, 300-302].

A second approach was designed to constrain the deployment of nuclear weapons, particularly tactical nuclear weapons, in NNWS. Yugoslavia, joined by Ghana, Nepal, Nigeria and Romania, submitted a draft resolution which

1. *Invites* the nuclear-weapon States Party to the Treaty to initiate, as soon as possible but not later than the end of 1976, negotiations on the conclusion of a treaty on the withdrawal from the territories of the non-nuclear-weapon States Party to the Treaty of all nuclear-weapon delivery systems, especially tactical nuclear weapons;
2. *Requests* the nuclear-weapon States Party to the Treaty to immediately discontinue further deployment of all types of tactical and other nuclear-weapon-delivery systems within the territories of the non-nuclear-weapon States Party to the Treaty and to simultaneously commence with their gradual withdrawal pending the entry into force of the aforementioned treaty;
3. *Invites* also the non-nuclear-weapon States Party to the Treaty on whose territories, waterways or air space the nuclear-weapon delivery systems are deployed not to allow the use or threat of use of nuclear weapons against other non-nuclear-weapon States Party to the Treaty[281, 284, 285, 296, 302, 303].

The draft resolution thus sought to limit and ultimately reverse the locational proliferation of tactical nuclear weapons which might be used on the territories of NNWS and/or against other NNWS in local wars. In the words of Ambassador Mihajlović of Yugoslavia:

The nuclear arms race had led to an unrestricted proliferation of tactical nuclear weapons and their deployment in areas which, at the time of entry into force of the [Non Proliferation] Treaty, had been free of such weapons.... The development of new nuclear explosives and of small yield nuclear warheads, intended for use in limited and local wars, had tended to facilitate the proliferation of tactical nuclear weapons and their stockpiling on the territories of foreign non-nuclear-weapon States. Such a proliferation seriously jeopardized the objectives of the Treaty and, in fact, obscured the true meaning of horizontal proliferation of nuclear weapons.

In his delegation's opinion, the deployment of nuclear weapons in the territories of non-nuclear-weapon States and the training of allied armed forces in their use represented an indirect nuclearization of those countries, which was incompatible with the spirit and objectives of the Treaty.... While not minimizing the fact that vital national interests of the nuclear-weapon States and their allies were involved in the SALT negotiations, the sponsors wished to stress that the vital interests of many non-nuclear-weapon States were directly threatened by the extensive proliferation of tactical nuclear weapons in many sensitive regions of the world[296].

A third approach supported the establishment of nuclear-free zones in various regions. In a draft resolution on Article VII of the NPT introduced by Iran, the parties to the NPT were asked to cooperate with states in regions which decide to establish nuclear-weapon-free zones. The NWS parties were also urged "to undertake a solemn obligation never to use or threaten to use nuclear weapons against countries which have become Parties to and are fully bound by the provisions of such regional arrangements"[276, 279-284, 286, 293, 295, 304].

Many NNWS had, during the Review Conference, urged that nuclear-free zones be established in such regions as the Balkans, the Middle East, Africa and South Asia, and that the NWS undertake in advance to respect them. Such zones were viewed as either supplements to universal negative security guarantees or substitutes for them should such guarantees prove impossible to implement on a universal basis.

The NWS expressed no interest in these approaches. Indeed, the USSR explicitly dismissed all three. It criticized Additional Protocol III on security assurances as imposing obligations on only a few NWS, claiming that the participation of all NWS would clearly be required if such undertakings were to succeed. The USSR viewed the draft resolution constraining the deployment of tactical nuclear weapons as falling outside the competence of the Review Conference and as possibly counter-productive for other negotiations — presumably for those on Mutual Force Reductions in Central Europe. Finally, the USSR deemed consideration of nuclear-free zones inappropriate for the Review Conference, since an Ad Hoc Group of Governmental Experts at the CCD was then initiating a "Comprehensive Study of the Question of Nuclear Weapon Free Zones in all its aspects". The USSR also stressed that any nuclear-free zone created in the future must "genuinely transform" the states involved into zones completely free of nuclear weapons and that such agreements must "exclude any loopholes for violating the non-nuclear status of the zones": the position upon which the USSR has traditionally opposed the existing Treaty of Tlatelolco[270, 305, 306].

The USA, while less explicitly negative, also made its lack of interest clear. In referring to security guarantees in general, Ambassador Klein concluded that

> in considering the usefulness of any proposals on the strengthening of the security of non-nuclear-weapon States, members should be guided not by any abstract concept concerning a balance of obligations under the Treaty but rather by the desire to pursue those measures that could suitably and effectively meet the legitimate and often pressing security concerns of those States[296].

And in specific reference to nuclear-free zones, the USA stressed that such zones should not disturb necessary security arrangements and that

> each nuclear free zone proposal must be judged on its own merits to determine whether the provision of specific security assurances would be likely to have a favourable effect. Moreover, we do not believe it would be realistic to expect

nuclear-weapon States to make implied commitments to provide such assurances before the scope and content of any nuclear-free zone arrangement are worked out[305].

In sum, the NWS felt no pressing need to respond further to the demands of NNWS for more forthcoming security guarantees, whether negative or positive.

Given this impasse on the issue of what security guarantees might be effectively related to the NPT, Conference President Thorsson only incorporated a few general points in the draft Final Declaration reviewing Article VII and the Security of Non-Nuclear Weapon States. First, the Conference "underlines the importance of adherence to the Treaty by non-nuclear-weapon States as the best means of reassuring one another" and of "strengthening their mutual security". Second, the Conference notes the intention of the depositary states to honour their commitments under Security Council Resolution 255. Third, the Conference recognizes the establishment of nuclear-weapon-free zones as an effective means of curbing proliferation and the need for cooperation by NWS. It cited the desire of "a considerable number of delegations" — that is, the NNWS — "that nuclear-weapon States should provide, in an appropriate manner, binding security assurances to those States which become fully bound by the provisions of such regional arrangements". Finally, the Conference urged the NWS parties to "ensure the security of all non-nuclear-weapon States Parties". To this end, the Conference urges all states, both NWS and NNWS, "to refrain, in accordance with the Charter of the United Nations, from the threat or the use of force in relations between states, involving either nuclear or non-nuclear weapons"[268].

Again, at the insistence of Ambassador Garcia Robles of Mexico who spoke on behalf of the group of 77 NNWS at the Review Conference, the texts of the Draft Resolutions recommending Additional Protocol III on Security Assurances, constraining tactical nuclear weapons and endorsing nuclear-free zones were annexed to the Final Declaration of the Review Conference[303, 304, 307]. Again, they served as mute testimony to the differences between the NWS and the NNWS on what constitutes a comprehensive and equitable security régime which might minimize the proliferation of nuclear weapons.

III. Future review of the NPT

The final Declaration stipulated that the parties to the treaty participating in the Review Conference "propose to the Depositary Governments that a second conference to review the operation of the Treaty be convened in 1980" in accordance with the procedures for convening periodic Review

Conferences under Article VIII of the NPT. It also invited participating states to request the Secretary-General of the United Nations to include "the implementation of the conclusions of the first Review Conference of the Parties to the Treaty on the Non Proliferation of Nuclear Weapons" on the provisional agenda of its thirty-first (1976) and thirty-third (1978) sessions.

Such future reviews may well continue to reveal the same profound differences as those in the 1975 Review Conference between the NWS and the NNWS concerning the type of arms limitation and security régime in which the NPT should be embedded. There had, in fact, been no change in position since the negotiations on the NPT. The 1975 Review Conference made no concrete proposals to spur nuclear arms limitation by the NWS or to promote the security of the NNWS. Clearly, the NWS parties continue to view the NPT as a single collateral measure of arms limitation primarily designed to constrain the horizontal proliferation of nuclear weapons. Equally clear, most NNWS, by contrast, are convinced that the NPT must be amplified by a comprehensive and counterdiscriminatory arms limitation and security régime which will both limit the nuclear weapon capabilities of existing NWS parties and better meet the various security objectives of the NNWS parties. The 1975 Review Conference of the NPT did nothing to breach this gap in perceptions and expectations.

5. Conclusion

This book explores the nature of the relationship between the arms limitation and security policies of the major states — particularly the NWS — and the acquisition of independent nuclear weapons capabilities by additional states. First, chapter 1 asserts that many NNWS are inexorably acquiring the technical and industrial capabilities required to develop and deploy nuclear weapons if they should perceive it in their military and/or political interests to do so. This process can be retarded through the adoption of varying international policies governing the export of nuclear fuel cycle facilities and international safeguard systems for such facilities, but it cannot be stopped. Thus, if one is interested in constraining nuclear proliferation, one must identify a set of long-term, mutually compatible arms limitation and security policies which might satisfy various objectives of NNWS, thereby minimizing their intentions to acquire independent nuclear weapon capabilities and collectively minimizing future nuclear proliferation.

Turning to the design of such an arms limitation and security régime, chapter 2 attempts to show that the modified Low Posture Doctrine would be the most effective comprehensive arms limitation and security régime the NWS could adopt in satisfying the various policy objectives of the NNWS and therefore minimizing the future proliferation of nuclear weapons. Chapters 3 and 4 attempt to show that much of the argument between the NWS and the NNWS during negotiations on the NPT from 1965 to 1968 and again at the Review Conference of the NPT in 1975 concerning the nature of the relationship between the arms limitation and security policies of the NWS and the nuclear decisions of the NNWS parallels the argument between advocates of the High Posture Doctrine and the Low Posture Doctrine, respectively. Specifically, these chapters suggest that the arms limitation and security policies which the NNWS have been seeking from the NWS are a close approximation of the obligations demanded of them under a modified Low Posture Doctrine.

As of the end of the Review Conference in 1975, these demands of the NNWS remained largely unmet. Recently, the new US administration of President Carter has made several moves consistent with the modified Low Posture Doctrine. It has renewed emphasis upon a comprehensive test ban and substantial reductions of strategic nuclear delivery vehicles in negotiations on strategic arms limitation after SALT II. It has enunciated certain declaratory and deployment policies, such as the planned withdrawal of theatre nuclear weapons from the Republic of Korea and the recent undertaking at the UN Special Session on Disarmament that it

> will not use nuclear weapons against any non-nuclear-weapons state party to the NPT (Non-proliferation Treaty) or any comparable internationally binding commitment not to acquire nuclear explosive devices, except in the case of an attack on the United States, its territories or armed forces, or its allies, by such a state allied to a nuclear-weapons state or associated with a nuclear-weapons state in carrying out or sustaining the attack[308].

These somewhat constrain US reliance upon nuclear weapons in the conduct of its foreign policy. However, they are at best only initial departures in what must be a substantial policy shift, subscribed to initially by both major NWS and ultimately by all NWS, if it is to constitute the comprehensive arms limitation and security régime demanded by NNWS and elaborated in the strategic literature as the modified Low Posture Doctrine.

The arms limitation and security policies demanded of the NWS under a modified Low Posture Doctrine impose obligations which will necessitate their making hard choices, adopting new norms of international behaviour, and accepting reduced freedom of action in international politics. However, if the logic of the modified Low Posture Doctrine is persuasive and if the NWS, as the major powers in the current international security system, are serious about the objective of minimizing future nuclear proliferation, the NWS must be prepared to undertake these obligations and to bear their costs. Conversely, continued adherence by the NWS to some approximation of the High Posture Doctrine will indicate that the NWS are not truly serious about minimizing future nuclear proliferation, and that proliferation will in all likelihood proceed. To quote Leonard Beaton on this final point:

> The future is therefore one of major political choice, not of technical capacity. If the major powers choose to create a structure which will effectively prevent proliferation over a long period of time, they must in the process change the facts of power. Weapons of mass destruction which seem to give their sovereign independence a secure future will force them to sink that independence in wider arrangements. If they choose to do nothing and leave others to take what decisions they must, the world around them will steadily become much less tolerable. Those who control the decisive weapons would obviously like to go on as we now are; but that is the one choice which is not open to them[34f].

References

1. Nuclear Energy Policy Study Group, *Nuclear Power: Issues and Choices,* sponsored by the Ford Foundation (Ballinger Publishing Co., Cambridge, Mass., 1977).
2. *World Armaments and Disarmament, SIPRI Yearbook 1977* (Almqvist & Wiksell, Stockholm, 1977, Stockholm International Peace Research Institute).
   (a) —, pp. 6-14.
   (b) —, appendix 1A.
   (c) —, chapter 2.
3. Barnaby, C.F., 'How states can go nuclear', *Annals of the Academy of Political and Social Sciences,* Vol. 430, March 1977, pp. 29-43.
4. Taylor, T. B. and Willrich, M., *Nuclear Theft: Risks and Safeguards* (Ballinger Publishing Co., Cambridge, Mass., 1974), pp. 59-76.
5. Bader, W. B., *The United States and the Spread of Nuclear Weapons* (Pegasus, New York, 1968), p. 92.
6. Quester, G., 'Some conceptual problems in nuclear proliferation', *American Political Science Review,* Vol. 66, No. 2, June 1972, p. 491.
7. Jensen, L., *Return From the Nuclear Brink* (D. C. Heath & Co., Lexington, Mass., 1974).
   (a) —, p. 84.
   (b) —, appendix C.
   (c) —, p. 79.
   (d) —, p. 30.
   (e) —, p. 234.
   (f) —, pp. 2, 25.
8. Dror, Y., 'Small powers' nuclear policy: research methodology and exploratory analysis', *Jerusalem Journal of International Relations,* Vol. 1, No. 1, Autumn 1975, pp. 30, 39, 45.
9. Marks, A., ed., *NPT: Paradoxes and Problems* (Arms Control Association, Washington D.C., 1975).
   (a) —, Goldblat, J., 'The Indian nuclear test and the NPT', p. 26
   (b) —, Smart, I., 'Non-Proliferation Treaty: status and prospects', pp. 26, 28-29, 41.
10. *World Armaments and Disarmament, SIPRI Yearbook 1975* (Almqvist & Wiksell, Stockholm, 1975, Stockholm International Peace Research Institute), pp. 22, 26, 521, 524-25.
11. Lodgaard, S., 'Reviewing the Non-Proliferation Treaty: status and prospects', *Instant Research on Peace and Violence*, Vol. 5, No. 1, 1975, p. 9.
12. Wadlow, R. V. L., 'The Nuclear Non-Proliferation Treaty Review Conference: May 1975', *Instant Research on Peace and Violence,* Vol. 5, No. 1, 1975, p. 3.

13. Lee, B. W., 'Korea's experiences in implementing a nuclear power programme', *International Atomic Energy Bulletin,* Vol. 17, No. 1, February 1975, p. 33.
14. Woite, G., 'The potential role of nuclear power in developing countries', *International Atomic Energy Bulletin,* Vol. 17, No. 1, June 1975, p. 30.
15. Hammond, S.A. *et al.*, 'Manpower requirements for future nuclear programmes', *International Atomic Energy Bulletin,* Vol. 17, No. 4, August 1975, p. 14.
16. Cairo, A., 'IAEA technical co-operation activities: Europe and the Middle East', *International Atomic Energy Bulletin,* Vol. 17, No. 4, August 1975, pp. 38-45.
17. *Nuclear Proliferation: Future U.S. Foreign Policy Implications,* Hearings before the Subcommittee on International Security and Scientific Affairs of the House International Relations Committee, 94th Congress, 1st Session (US Government Printing Office, Washington D.C., 1975), pp. 390-401.
18. Dunn, L. A., 'Nuclear grey marketeering', *International Security,* Vol. 1, No. 3, Winter 1977, pp. 107-18.
19. *New York Times,* 15 March 1976.
20. Baker, S. J., 'The international political economy of proliferation', ed. D. Carleton and C. Schaerf, *Arms Control and Technical Innovation* (Croom Helm, London, 1977).
21. Baker, S. J., 'Monopoly or cartel?' *Foreign Policy,* No. 23, Summer 1976, pp. 202-20.
22. *Safeguards Against Nuclear Proliferation* (Almqvist & Wiksell, Stockholm, 1975, Stockholm International Peace Research Institute).
23. Greenwood, T. *et al., Nuclear Proliferation* (McGraw-Hill Book Co., New York, 1977).
24. Wohlstetter, A., 'Spreading the bomb without quite breaking the rules', *Foreign Policy*, No. 25, Winter 1976-77, p. 88.
25. *Moving Toward Life in a Nuclear Armed Crowd?,* Final Report prepared for the US Arms Control and Disarmament Agency by Pan Heuristics, ACDA/PAB-263, PH 76-04-389-14, revised 22 April 1976.
26. Holst, J. J., ed., *Security, Order and the Bomb* (Universitetsförlaget, Oslo, 1972).
(a) —, Holst, J. J., 'Perspectives on post-NPT proliferation issues: an introduction', pp. 9, 12.
(b) —, Subrahmanyam, K., 'The role of nuclear weapons: an Indian perspective', p. 134.
(c) —, Nerlich, U., 'Nuclear weapons and European politics: some structural interdependencies', pp. 74-75, 78-83, 91.
(d) —, Hoag, M., 'One American perspective on nuclear guarantees,

proliferation and related alliance diplomacy', *passim*, pp. 156, 158, 162, 164-66, 182.

27. Bull, H., 'Rethinking non-proliferation', *International Affairs,* Vol. 51, No. 2, 1 April 1975, pp. 15, 175-80, 187-89.
28. Bobrow, D. B., *Technology-Related International Outcomes: R and D Strategies to Induce Sound Public Policy* (Pittsburgh, Pa., 1974, International Studies Association), *passim*.
29. Ra'anan, U., 'Some political perspectives concerning the US-Soviet strategic balance', ed. G. Kemp *et al., The Superpowers in a Multinuclear World* (D. C. Heath & Co., Lexington, Mass., 1974).
    (a) —, p. 20.
30. Rosen, S. J., 'Nuclearization and stability in the Middle East', Brandeis University, 1975 (unpublished research paper).
31. Sandoval, R. R., 'Consider the porcupine: another view of nuclear proliferation', *Bulletin of Atomic Scientists*, Vol. 32, No. 5, May 1975.
32. *New York Times,* 14 November 1974 (speech by Yasur Arafat).
33. Dror, Y., *Crazy States* (D. C. Heath & Co., Lexington, Mass., 1971).
34. Beaton, L., *Must the Bomb Spread?* (Penguin Books, London, 1966).
    (a) —, pp. 23, 131.
    (b) —, p. 129.
    (c) —, p. 118.
    (d) —, pp. 119, 126-27.
    (e) —, pp. 123-27.
    (f) —, pp. 132-33.
35. Buchan, A., ed., *A World of Nuclear Powers?* (Columbia University Press, New York, 1966, The American Assembly).
    (a) —, Hoffmann, S., 'Nuclear proliferation and world politics', pp. 89-121.
    (b) —, Buchan, A., 'Introduction', pp. 1-2.
    (c) —, Chalfont, Lord, 'Alternatives to proliferation: inhibition by agreement', p. 125.
36. Rosecrance, R., ed., *The Future of the International Strategic System* (Chandler Publishing Co., San Francisco, 1972).
    (a) —, Rosecrance, R., 'Introduction', pp. 1-9
    (b) —, Rosecrance, R., 'Reward, punishment, and the future', pp. 175-84.
    (c) —, Quester, G., 'The politics of twenty nuclear powers', pp. 56-77.
    (d) —, Hoag, M., 'Superpower strategic postures for a multipolar world', pp. 41-42, 48.
37. Haskel, B., 'Disparities, strategies, and opportunity costs: the example of Scandinavian economic market negotiations', *International Studies Quarterly,* Vol. 18, No.1, March 1974, p. 27.

38. Myrdal, A., 'The game of disarmament', *Impact,* Vol. 22, No. 3, July-September 1972, pp. 38, 228.
39. Myrdal, A., *The Game of Disarmament* (Pantheon, New York, 1976).
40. Tucker, R. C. *et al., Proposals For No First Use of Nuclear Weapons: Pros and Cons* (Princeton, Center for International Studies, 1963).
    (a) — , Falk, R. A., 'Some thoughts in support of a no first use proposal', pp. 40-42, 54-63, 144.
    (b) — , Tucker, R. C., 'No first use of nuclear weapons', pp. 12-16.
41. Falk, R.A., *Legal Order in a Violent World* (Princeton University Press, Princeton, 1968), p. 371.
42. Beaton, L., *The Reform of Power* (Viking Press, New York, 1972).
    (a) —, p. 12.
    (b) —, pp. 204-15.
    (c) —, pp. 120-21, 129-30.
    (d) —, ch. 5, 7.
43. Bull, H., 'The scope for super-power agreements', *Arms Control and National Security,* Vol. 1, 1969, pp. 6, 7, 14.
44. *Safeguarding the Atom*, Report of a Policy Panel established by UNA/USA and a Report of a Policy Panel established by the Association for the UN in the USSR (UNA/USA Publishing House, New York, 1972).
45. Bull, H., 'The twenty years crisis thirty years on', *International Journal,* Vol. 24, No. 4, Autumn 1969, pp. 637-38.
46. Mendlovitz, S.H., *On the Creation of a Just World Order* (Free Press, New York, 1975).
47. Singer, J. D., 'The level of analysis problem in international relations', ed. R. Falk and S. Mendlovitz, *The Strategy of World Order Vol. 1: Toward a Theory of War Prevention* (New York, World Law Fund, 1961).
48. Wight, M., 'Western values in international relations', ed. H. Butterfield and M. Wight, *Diplomatic Investigations* (Allen & Unwin, London, 1966), pp. 102 ff.
49. Quester, G., *The Politics of Nuclear Proliferation* (Johns Hopkins University Press, Baltimore, 1973).
    (a) —, pp. 2, 235.
50. Lawrence, R. M. and Larus, J., eds:, *Nuclear Proliferation Phase II* (University Press of Kansas, Lawrence, Kans., 1974).
51. Willrich, M., *The Non-Proliferation Treaty: Framework for Nuclear Arms Control* (The Michie Co., Charlottesville, Va., 1969)
    (a) —, p. 9.
    (b) —, p. 171.
    (c) —, pp. 148, 155-68.

(d) —, p. 46.
(e) —, p. 100.
52. Epstein, W., 'Why states go — and don't go — nuclear', *Annals of the Academy of Political and Social Sciences*, No. 430, March 1977, pp. 16-28.
53. Quester, G. H., 'Reducing the incentives to proliferation', *Annals of the Academy of Political and Social Sciences,* No. 430, March 1977, pp. 70-81.
54. Betts, R. K., 'Paranoids, pygmies, pariahs, and nonproliferation', *Foreign Policy,* Spring 1977, pp. 157-83.
55. Allison, G., *Essence of Decision* (Little, Brown, Boston, 1971).
56. *FY 1976 and FY 197T Annual Defense Department Report.*
(a) —, chapter 3, p. 2.
(b) —, chapter 2, p. 6.
(c) —, chapter 2, pp. 2-4.
(d) —, chapter 1, p. 13.
(e) —, chapter 2, p. 7.
57. Kemp, G., *Nuclear Forces for Medium Powers, Parts I and II,* Adelphi Papers Nos. 106 and 107 (London, International Institute for Strategic Studies, 1974).
58. Vital, D., *The Inequality of States* (Clarendon Press, Oxford, 1967), chapter 9.
59. Cox, R. W. and Jacobson, H. K., *The Anatomy of Influence* (Yale University Press, New Haven, 1973), appendix A.
60. Maddox, J., *Prospects for Nuclear Proliferation,* Adelphi Paper No. 113 (London, International Institute for Strategic Studies, 1975).
(a) —, pp. 19-20.
(b) —, pp. 5-6, 31-33.
(c) —, pp. 7-8, 21-22, 32.
61. Bloomfield, L., 'Nuclear spread and world order', *Foreign Affairs,* Vol. 53, No. 4, July 1975.
(a) —, pp. 746, 48.
62. Dore, R. P., 'The prestige factor in international affairs', *International Affairs,* Vol. 51, No. 2, April 1975, p. 202.
63. Issacs, H., 'Nationality: "the end of the road" ', *Foreign Affairs,* Vol. 53, No. 3, April 1975.
64. *The Near-Nuclear Countries and the Non-Proliferation Treaty* (Almqvist & Wiksell, Stockholm, 1972, Stockholm International Peace Research Institute), p. 76.
65. Dougherty, J. and Lehman, J. F., eds., *Arms Control for the Late Sixties* (P. Van Nostrand Co., Princeton, 1967).
(a) —, Bull, H., 'The role of nuclear powers in the management of nuclear proliferation', pp. 143, 145, 148, 150.

(b) —, Schlesinger, J., 'The strategic consequences of nuclear proliferation', pp. 174-84.
(c) —, Kahn, H., 'Suggestions for long-term anti-nuclear policies', pp. 166-68.
(d) —, Silard, J., 'The coming international nuclear security treaty', pp. 186, 191.

66. Singer, M., 'A non-utopian, non-nuclear future world', *Arms Control and Disarmament,* Vol. 1, 1968, *passim*, pp. 80-81, 93.
67. Hahn, W., 'Nuclear proliferation', *Strategic Review,* Vol. 3, No. 1, Winter 1975.
    (a) —, p. 24.
    (b) —, pp. 23-24.
    (c) —, p. 22.
68. Frye, A., 'How to ban the bomb: sell it', *New York Times Magazine,* 11 January 1976.
    (a) —, p. 11.
    (b) —, pp. 76-78.
69. *FY 1975 and FY 197T Annual Defense Department Report.*
    (a) —, chapters 1-2, *passim*.
    (b) —, p. 41.
    (c) —, pp. 27, 36, 38.
70. Frye, A., *A Responsible Congress* (McGraw-Hill Book Co., New York, 1975), chapters 3, 4 and 5.
71. Greenwood, T. and Nacht, M., 'The new nuclear debate: sense or nonsense?' *Foreign Affairs,* Vol. 52, No. 4, July 1974, p. 775.
72. *New York Times*, 26 June 1975, p. 8.
73. Willrich, M. and Boskey, B., eds., *Nuclear Proliferation: Prospects for Control* (Dunellen Publishing Co., New York, 1970).
    (a) —, Coffey, J. I., 'Threat reassurance and nuclear proliferation', pp. 122, 131-32.
    (b) —, Falk, R., 'Renunciation of nuclear weapons use', pp. 133 ff.
74. Falk, R., 'Arms control, foreign policy and global reform', *Daedalus*, Vol. 104, No. 3, Summer 1975.
75. Falk, R., 'Arms control, foreign policy and global autonomy', *Daedalus,* Vol. 104, No. 3, Summer 1975.
76. Doty, P., 'Strategic arms limitation after SALT I', *Daedalus,* Vol. 104, No. 3, Summer 1975, p. 67.
77. Brooks, H., 'The military innovation system and the qualitative arms race', *Daedalus*, Vol. 104, No. 3, Summer 1975.
78. Vital, G., 'The problems of guarantees', ed. C. F. Barnaby, *Preventing the Spread of Nuclear Weapons* (Souvenir Press, London, 1969), pp. 147, 149.

79. *NPT: The Review Conference and Beyond,* Report of a Policy Panel established by the UNA/USA (UNA/USA Publishing House, New York, 1975).
80. *The United Nations and Disarmament 1945-70,* UN Department of Political and Security Affairs (New York, 1970).
    (a) —, pp. 371-2.
    (b) —, pp. 267-70.
    (c) —, pp. 288, 307-309.
81. Ullman, R., 'No first use of nuclear weapons', *Foreign Affairs,* Vol. 50, No. 4, July 1972, pp. 673, 681.
82. Willrich, M., 'No first use of nuclear weapons — an assessment', *Orbis,* Vol. 9, No. 2, Summer 1965, p. 309.
83. *SIPRI Yearbook of World Armaments and Disarmament 1969/70* (Almqvist & Wiksell, Stockholm, 1970, Stockholm International Peace Research Institute).
84. Wiesner, J., 'Comprehensive arms-limitation systems', ed. J. Wiesner, *Where Science and Politics Meet* (McGraw-Hill Book Co., New York, 1965).
85. Bundy, McG., 'To cap the volcano', *Foreign Affairs,* Vol. 48, No. 1, October 1969.
86. Maddox, J., 'No need for NPT gloom', *New Scientist,* June 1975.
87. Official Records of the UN General Assembly, Thirteenth Session, Annexes, Agenda items 64, 70 and 72. UN document A/C.1/L.206.
88. Disarmament Conference document ENDC/120.
89. Disarmament Conference document ENDC/123.
90. Foster, W. C., 'New directions in arms control', *Foreign Affairs,* Vol. 43, No. 3, July 1965, pp. 587-601.
91. Disarmament Conference document DC/225, paragraph 2(c).
92. *SIPRI Yearbook of World Armaments and Disarmament, 1968/69* (Almqvist & Wiksell, Stockholm, 1970, Stockholm International Peace Research Institute), pp. 156-59.
93. Disarmament Conference document ENDC/162.
94. UN document A/5976.
95. UN document A/C.1/SR.1355.
96. UN document A/C.1/SR.1373.
97. Disarmament Conference document ENDC/PV.228.
98. Disarmament Conference document ENDC/PV.299.
99. Disarmament Conference document ENDC/PV.262.
100. Disarmament Conference document ENDC/PV.265.
101. Disarmament Conference document ENDC/PV.264.
102. Disarmament Conference document ENDC/PV.286.
103. Disarmament Conference document ENDC/PV.268.

104. Disarmament Conference document ENDC/PV.241.
105. Disarmament Conference document ENDC/PV.244.
106. Disarmament Conference document ENDC/PV.245.
107. Disarmament Conference document ENDC/PV.288.
108. Disarmament Conference document ENDC/PV.273.
109. Disarmament Conference document ENDC/165.
110. Disarmament Conference document ENDC/PV.246.
111. Disarmament Commission Official Records, 75th Meeting, p. 4.
112. Disarmament Conference document ENDC/PV.240.
113. Disarmament Conference document ENDC/PV.223.
114. Disarmament Conference document ENDC/PV.232.
115. Disarmament Conference document ENDC/PV.298.
116. Disarmament Conference document ENDC/PV.263.
117. Disarmament Conference document ENDC/PV.243.
118. Disarmament Conference document ENDC/PV.281.
119. Disarmament Conference document ENDC/PV.300.
120. Disarmament Conference document ENDC/PV.242.
121. Disarmament Conference document ENDC/PV.250.
122. Disarmament Conference document ENDC/PV.318.
123. UN document A/C.1/SR.1359.
124. Disarmament Conference document ENDC/PV.229.
125. Disarmament Conference document ENDC/PV.235.
126. Disarmament Conference document ENDC/158.
127. Disarmament Conference document ENDC/178.
128. Disarmament Conference document ENDC/PV.289.
129. Disarmament Conference document ENDC/PV.303.
130. Disarmament Conference document ENDC/PV.304.
131. Disarmament Conference document ENDC/PV.293.
132. Disarmament Conference document ENDC/192.
133. Disarmament Conference document ENDC/193.
134. Disarmament Conference document ENDC/PV.280.
135. Disarmament Conference document ENDC/PV.330.
136. Disarmament Conference document ENDC/PV.369.
137. Disarmament Conference document ENDC/PV.313.
138. Disarmament Conference document ENDC/PV.339.
139. Disarmament Conference document ENDC/PV.366.
140. Disarmament Conference document ENDC/PV.302.
141. Disarmament Conference document ENDC/PV.337.
142. Disarmament Conference document ENDC/PV.334.
143. Disarmament Conference document ENDC/PV.340.

144. Disarmament Conference document ENDC/PV.342.
145. Disarmament Conference document ENDC/PV.335.
146. Disarmament Conference document ENDC/PV.331.
147. Disarmament Conference document ENDC/PV.333.
148. Disarmament Conference document ENDC/PV.336.
149. Disarmament Conference document ENDC/199.
150. Disarmament Conference document ENDC/196.
151. Disarmament Conference document ENDC/201.
152. UN document A/C.1/SR.1436.
153. Disarmament Conference document ENDC/PV.370.
154. Disarmament Conference document ENDC/PV.297.
155. Disarmament Conference document ENDC/PV.363.
156. Disarmament Conference document ENDC/PV.338.
157. Disarmament Conference document ENDC/PV.356.
158. Disarmament Conference document ENDC/PV.350.
159. UN document A/C.1/PV.1553.
160. UN document A/C.1/PV.1555.
161. Disarmament Conference document ENDC/192/Rev. 1.
162. Disarmament Conference document ENDC/193/Rev. 1.
163. Disarmament Conference document ENDC/PV.357.
164. Disarmament Conference document ENDC/PV.362.
165. Disarmament Conference document ENDC/PV.368.
166. Disarmament Conference document ENDC/PV.361.
167. Disarmament Conference document ENDC/PV.358.
168. Disarmament Conference document ENDC/PV.376.
169. Disarmament Conference document ENDC/223/Rev. 1.
170. Disarmament Conference document ENDC/201/Rev. 2.
171. Disarmament Conference document ENDC/215.
172. Disarmament Conference document ENDC/216.
173. Disarmament Conference document ENDC/PV.364.
174. Disarmament Conference document ENDC/PV.373.
175. Disarmament Conference document ENDC/PV.365.
176. Disarmament Conference document ENDC/PV.367.
177. Disarmament Conference document ENDC/PV.371.
178. Disarmament Conference document ENDC/224, Annex A.
179. Disarmament Conference document ENDC/PV.378.
180. UN document A/C.1/PV.1568.
181. UN document A/C.1/PV.1571.
182. UN document A/C.1/PV.1573.
183. UN document A/C.1/PV.1556.

184. UN document A/C.1/PV.1577.
185. Remarks by US President on the signing of the Nonproliferation Treaty, *US Department of State Bulletin,* 22 July 1968, pp. 85-87.
186. UN document A/C.1/PV.1569.
187. UN document A/C.1/PV.1572.
188. UN document A/C.1/PV.1564.
189. UN document A/C.1/PV.1560.
190. UN document A/C.1/PV.1565.
191. UN document A/C.1/PV.1566.
192. UN document A/C.1/PV.1576.
193. UN document A/C.1/PV.1567.
194. UN document A/C.1/PV.1561.
195. UN document A/C.1/PV.1563.
196. UN document A/C.1/PV.1562.
197. UN document A/C.1/PV.1580.
198. UN document A/C.1/PV.1574.
199. UN document A/C.1/PV.1575.
200. UN document A/C.1/PV.1559.
201. UN document A/C.1/PV.1579.
202. UN document A/C.1/L.421/Rev. 2/Add. 1, 31 May 1968.
203. Address by US President, *US Department of State Bulletin,* 2 November 1964, p. 613.
204. Disarmament Conference document ENDC/PV.239.
205. Press conference remarks by Secretary of State Rusk, *US Department of State Bulletin,* 6 June 1966, pp. 884-85.
206. Disarmament Conference document ENDC/167.
207. Disarmament Conference document ENDC/PV.267.
208. Larson, E., *Disarmament and Soviet Policy, 1964-1968.* (Prentice Hall Inc., Englewood Cliffs, N.J., 1969), pp. 153-54.
209. Disarmament Conference document ENDC/PV.261.
210. Disarmament Conference document ENDC/PV.282.
211. Disarmament Conference document ENDC/PV.319.
212. Disarmament Conference document ENDC/PV.292.
213. Disarmament Conference document ENDC/PV.294.
214. Disarmament Conference document ENDC/PV.310.
215. Disarmament Conference document ENDC/180.
216. Disarmament Conference document ENDC/PV.274.
217. Disarmament Conference document ENDC/PV.295.
218. UN document A/C.1/SR.1365.
219. Disarmament Conference document ENDC/PV.237.

220. Disarmament Conference document ENDC/PV.222.
221. Extract from News Conference Remarks by Indian External Affairs Minister Chagla, 27 April 1967, *Documents on Disarmament, 1967* (US Government Printing Office, Washington D.C., 1968, US Arms Control and Disarmament Agency).
(a) —, p. 204.
(b) —, p. 206.
222. Disarmament Conference document ENDC/PV.323.
223. UN document A/PV.1562.
224. Disarmament Conference document ENDC/PV.329.
225. Disarmament Conference document ENDC/PV.327.
226. Disarmament Conference document ENDC/PV.344.
227. Disarmament Conference document ENDC/204.
228. Disarmament Conference document ENDC/197.
229. Disarmament Conference document ENDC/202.
230. Disarmament Conference document ENDC/PV.351.
231. Disarmament Conference document ENDC/PV.346.
232. Memorandum from the Federal Republic of Germany to Other Governments, US Department of State, *Documents on Disarmament, 1968* (US Government Printing Office, Washington D.C., 1969, US Arms Control and Disarmament Agency), pp. 154-55.
233. Disarmament Conference document ENDC/220/Rev. 1.
234. Disarmament Conference document ENDC/222.
235. Disarmament Conference document ENDC/PV.375.
236. Disarmament Conference document ENDC/225.
237. UN document A/7072.
238. Disarmament Conference document DC/230.
239. Disarmament Conference document ENDC/378.
240. UN document A/C.1/PV.1582.
241. UN document S/PV.1433.
242. UN document A/C.1/PV.1570.
243. UN document A/C.1/PV.1366.
244. UN document A/C.1/PV.1578.
245. Disarmament Conference document ENDC/PV.224.
246. UN document A/C.1/SR.1363.
247. Disarmament Conference document ENDC/157.
248. Disarmament Conference document ENDC/PV.219.
249. Disarmament Conference document ENDC/PV.284.
250. Disarmament Conference document ENDC/PV.325.
251. Disarmament Conference document ENDC/PV.326.

252. Disarmament Conference document ENDC/203.
253. Disarmament Conference document ENDC/PV.341.
254. Disarmament Conference document ENDC/PV.348.
255. Disarmament Conference document ENDC/200/Rev. 1.
256. Disarmament Conference document ENDC/203/Rev. 1.
257. Disarmament Conference document ENDC/219.
258. Disarmament Conference document ENDC/201/Rev. 2.
259. Disarmament Conference document ENDC/218.
260. Disarmament Conference document ENDC/220.
261. Fletcher, G. P., 'The Presumption of Innocence in the Soviet Union', *UCLA Law Review,* Vol. 15, No. 4, 4 June 1968.
262. Disarmament Conference document ENDC/PV.307.
263. Disarmament Conference document ENDC/PV.296.
264. UN document A/C.1/1556.
265. *World Armaments and Disarmament, SIPRI Yearbook 1976* (Almqvist & Wiksell, Stockholm, 1976, Stockholm International Peace Research Institute), pp. 363-92.
266. Marwah, O. and Schulz, A., eds., *Nuclear Proliferation and the Near-Nuclear Countries* (Ballinger Publishing Co., Cambridge, Mass., 1975), pp. 301-13.
267. 'Working Paper on the Final Documents of the NPT Review Conference', NPT Preparatory Committee, 13 February 1975.
268. NPT Review Conference Document NPT/CONF/35/I/Annex I.
269. NPT Review Conference Document NPT/CONF/SR.2.
270. NPT Review Conference document NPT/CONF/C.1/SR.13.
271. NPT Review Conference document NPT/CONF/C.I/4.
272. NPT Review Conference document NPT/CONF/C.I/SR.4.
273. NPT Review Conference document NPT/CONF/C.I/SR.5.
274. NPT Review Conference document NPT/CONF/C.I/SR.2.
275. NPT Review Conference document NPT/CONF/C.I/SR.1.
276. NPT Review Conference document NPT/CONF/SR.3.
277. NPT Review Conference document NPT/CONF/C.I/7.
278. NPT Review Conference document NPT/CONF/C.I/9.
279. NPT Review Conference document NPT/CONF/SR.4.
280. NPT Review Conference document NPT/CONF/SR.5.
281. NPT Review Conference document NPT/CONF/SR.6.
282. NPT Review Conference document NPT/CONF/SR.7.
283. NPT Review Conference document NPT/CONF/SR.9.
284. NPT Review Conference document NPT/CONF/SR.10.
285. NPT Review Conference document NPT/CONF/C.I/SR.3.
286. NPT Review Conference document NPT/CONF/SR.8.

287. NPT Review Conference document NPT/CONF/26.
288. NPT Review Conference document NPT/CONF/17/Add. 1-4.
289. NPT Review Conference document NPT/CONF/18/Add. 1-3.
290. NPT Review Conference document NPT/CONF/17.
291. NPT Review Conference document NPT/CONF/18.
292. NPT Review Conference document NPT/CONF/C.I/8.
293. NPT Review Conference document NPT/CONF/C.I/SR.11.
294. NPT Review Conference document NPT/CONF/C.I/L.3.
295. NPT Review Conference document NPT/CONF/C.I/SR.7.
296. NPT Review Conference document NPT/CONF/C.I/SR.9.
297. NPT Review Conference document NPT/CONF/L.2/Rev. 1.
298. NPT Review Conference document NPT/CONF/L.3/Rev. 1.
299. NPT Review Conference document NPT/CONF/C.I/5/Add. 1.
300. NPT Review Conference document NPT/CONF/22.
301. NPT Review Conference document NPT/CONF/C.I/SR.6.
302. NPT Review Conference document NPT/CONF/C.I/SR.10.
303. NPT Review Conference document NPT/CONF/C.I/L.1.
304. NPT Review Conference document NPT/CONF/C.I/L.2.
305. NPT Review Conference document NPT/CONF/35/I/Annex II.
306. UN General Assembly, Resolution F (XXIX) 'Special Report of the CCD', A/1002/ADD (United Nations, New York, 1976).
307. NPT Review Conference Document NPT/CONF/L.4/Rev. 1.
308. *UN Special Session on Disarmament: Achievements*, Bureau of Public Affairs, US Department of State (US Government Printing Office, Washington, D.C., November 1978).

Appendix

SIPRI publications on related aspects of nuclear non-proliferation, 1973-78

SIPRI Yearbooks

World Armaments and Disarmament, SIPRI Yearbook 1973 (Almqvist & Wiksell, Stockholm, 1973, Stockholm International Peace Research Institute), pp. 422-25.

World Armaments and Disarmament, SIPRI Yearbook 1974 (Almqvist & Wiksell, Stockholm, 1974, Stockholm International Peace Research Institute), pp. 440-41; and appendix 13D, Protocols signed together with certain IAEA nuclear safeguards agreements.

World Armaments and Disarmament, SIPRI Yearbook 1975 (Almqvist & Wiksell, Stockholm, 1975, Stockholm International Peace Research Institute), chapter 2, Nuclear-weapon proliferation; pp. 493-99; appendix 15D, Official reactions to the first Indian nuclear explosion; appendix 15E, Status of NPT safeguards agreements with non-nuclear-weapon states, as of 31 January 1975; appendix 15F, Agreements providing for IAEA safeguards other than those in connection with the NPT, approved by the IAEA Board, as of 31 January 1975; and appendix 15G, Memorandum B attached to the letters from supplier countries, addressed in 1974 to the Director-General of the IAEA.

World Armaments and Disarmament, SIPRI Yearbook 1976 (Almqvist & Wiksell, Stockholm, 1976, Stockholm International Peace Research Institute), pp. 6-11; appendix 1B, The spread of nuclear power; chapter 9, section I, The Non-Proliferation Treaty; and appendix 9A, Final declaration of the Review Conference of the parties to the Treaty on the non-proliferation of nuclear weapons, 30 May 1975.

World Armaments and Disarmament, SIPRI Yearbook 1977 (Almqvist & Wiksell, Stockholm, 1977, Stockholm International Peace Research Institute), pp. 6-14; appendix 1A, The London Club; and chapter 2, The increase in international nuclear transactions.

World Armaments and Disarmament, SIPRI Yearbook 1978 (Taylor & Francis, London, 1978, Stockholm International Peace Research Institute), chapter 2, The nuclear fuel cycle and nuclear proliferation; and appendix 2A, Nuclear-export guidelines of the London Club.

Other SIPRI books

Nuclear Proliferation Problems (Almqvist & Wiksell, Stockholm, 1974, Stockholm International Peace Research Institute).

The Nuclear Age (Almqvist & Wiksell, Stockholm, 1975, Stockholm International Peace Research Institute).

Safeguards Against Nuclear Proliferation (Almqvist & Wiksell, Stockholm, 1975, Stockholm International Peace Research Institute).

Armaments and Disarmament in the Nuclear Age (Almqvist & Wiksell, Stockholm, 1975, Stockholm International Peace Research Institute).

Index

ABM (Anti-Ballistic Missile) 41, 56 *see also* Treaty
ASW (Anti-Submarine Warfare) 41
Afghanistan 96, 97, 111, 112, 114, 117
Africa 60, 103, 143
    Southern 15, 35 *see also* South Africa
Agreement
    'Hot Line' 77
    Measures to reduce risk of nuclear war (1971) 131
    no-first-use 42, 44, 46, 49, 54, 58, 70, 139
    Prevention of incidents on and over the high seas (1971) 131
    Prevention of nuclear war (1973) 131
    zone-of-peace 43
    see also SALT
Akwei, R.M. (*Ambassador*) 113
Albania 111, 112, 116, 118
Algeria 21, 22, 96, 97, 98, 114, 115, 116, 118
America, Latin, 15, 34, 103, 111
    Prohibition of Nuclear Weapons in 34 *see also* Treaty of Tlatelolco
Antarctica *see under* Treaty
Argentina 3, 8, 18, 21, 22, 96, 98, 110, 111, 112
Arms limitation 7ff., 14ff., 57, 69, 72, 76, 77ff., 86, 87, 92, 93, 96, 98, 105, 119, 120, 121, 122, 126, 127, 128, 129, 130, 131ff., 144, 145
Arms race 6, 60, 93, 96, 132
Asia
    East 15, 17
    South 15, 17, 60, 143
Australasia 15, 17
Australia 3, 17, 18, 21, 22, 32, 136
Austria 3, 16, 22, 96

Balkans 143
Barbados 117, 118
Beaton, L. 7, 12, 13, 39, 49, 50, 51, 54, 56, 63, 147
Belgium 3, 16, 22
Bergmann, E. 2
Biological weapons 46, 61
    convention (1971) 47, 131
Bloomfield, L. 65
Bolivia 135, 141
Brazil 3, 8, 18, 20, 21, 22, 63, 76, 82, 90, 91, 92, 93, 96, 98, 103, 104, 111, 112, 118, 122, 124
Buchan, A. 12
Budget, military; reduction of 131
Bulgaria 3, 16, 22, 94, 100
Bull, H. 5, 12, 13, 25, 26, 39, 44, 48, 49, 51, 54, 57, 64, 66
Burma 82, 85, 89, 98, 106, 111, 112, 123
Burns, E.L.M. (*Ambassador*) 86, 101, 107

CCD (Conference of the Committee on Disarmament) 47, 82, 136, 143
CTB (Comprehensive Test Ban) 30, 41, 61, 75, 84, 85, 94, 95, 97, 120, 132, 134, 135, 136, 137, 138, 146
Canada 3, 8, 22, 82, 86, 90, 94, 100, 104, 107, 123, 124, 125
Carter, J. (*President*) 146
Castaneda, J. (*Ambassador*) 90, 92
Ceylon 96, 97, 112, 116 *see also* Sri Lanka
Chalfont, *Lord* 79, 80, 81, 128
Chemical weapons 8, 46, 47, 61
    control of 131
Chen Pao 15
Chile 3, 19, 21, 22, 112, 118
China (People's Republic of) 2, 8, 11, 15, 16, 17, 20, 31, 32, 38, 48, 56, 58, 62, 63, 100, 105, 115, 116, 132
    nuclear test (1964) 24, 100
Cohen, S.B. 20
Colombia 112, 117, 118
Congo 113
Conventional arms control 8

Conventional attack 15-18, 31, 33, 48, 49, 55, 57-59, 67, 70
Counterforce capability *see* Nuclear weapons, first strike
'Crazy states' 18, 19, 21
Cuba 17, 19, 21, 22, 97, 98, 114, 115, 116, 118
Cyprus 96, 97, 98, 111, 112, 114
Czechoslovakia 3, 16, 22, 100

Dahomey 96, 111, 112, 113, 114, 115, 117
Damansky Island 15
Danieli, A.B. (*Ambassador*) 112
Deployment restrictions *see under* Nuclear weapons
Disarmament 77ff., 87, 92, 93, 96, 97, 98, 105, 106, 119, 120, 121, 122, 126, 127, 128, 129, 130, 131ff.
Dominican Republic 113
Doty, P. 40
Dror, Y. 6
Dual-capable delivery systems 40

ENDC (Eighteen Nation Disarmament Committee)/CCD 9, 75, 76, 77, 85, 86, 89, 90, 91, 95, 96, 98, 99, 100, 101, 102, 105, 106, 108, 111, 120, 123, 128, 129 *see also* CCD
Eastern Europe 7
Ecobesco, N. (*Ambassador*) 122
Ecuador 135, 141
Egypt 3, 8, 18, 21, 22, 82, 140
El Salvador 96, 112, 117
Environmental warfare
  prohibition of 131
  sea-bed 132
Ethiopia 46, 82, 85, 90, 94, 96, 103, 107, 110, 111, 124, 125
Euratom 87
Europe 15, 48, 49, 53, 55
  Central 49, 143
  —, Mutual Force Reductions 143
  Western 49
  see also Eastern Europe, Western European Community
Export of nuclear facilities 4

Falk, R. 10, 12, 39, 40, 44, 46, 48, 49, 51, 54, 55, 57, 64
Fanfani, A. 120
Fanfani Proposal (1965) 97, 120, 123
Fartash, M. (*Ambassador*) 133
Finland 3, 16, 22, 110
First-strike capability *see under* Nuclear weapons
Fisher, A. (*Ambassador*) 92, 119
Fissile material 2, 4, 126
  diversion of 4
  production 41, 79, 84, 97, 134
Flexible response (doctrine of) 16
Foster, W.C. (*Ambassador*) 79, 119, 121
France 2, 8, 11, 16, 19, 20, 31, 38, 62, 75, 115, 118, 132
Frye, A. 27, 28, 29, 30, 31, 32, 33, 34, 35, 36, 66

GCD (General and Complete Disarmament) 74, 76, 90, 91, 93, 94, 96, 129
GNP (Gross National Product) 16, 71
Geneva 121, 125, 129
Geneva Protocol (1925) 46, 47
German Democratic Republic 3, 16, 22
Germany (Federal Republic of) 3, 16, 20, 22, 53, 63, 70, 75, 101, 108, 124, 139
Ghana 97, 112, 113, 115, 116, 135, 140, 141, 142
Goldberg, A.J. (*Ambassador*) 95, 110, 128
Greece 3, 16, 18, 21, 22, 111, 112
Guarantee 99ff., 138, 144 *see also* Security
  no-first-use 31 *see also* Agreement
Guyana 96, 98, 111, 112

Hahn, W. 27
Hamilton, G. (*Ambassador*) 136
Haskel, B. 8
Herbicides 47
High Posture Doctrine *see under* Strategy for non-proliferation
Hoag, M. 27, 32, 39
Honduras 135
Hungary 3
Husain, M.A. (*Ambassador*) 97

IAEA (International Atomic Energy Agency) 60, 87, 127
Iklé, F.C. (*Ambassador*) 132

Imru, M. (*Ambassador*) 119
India 2, 8, 17, 28, 32, 38, 58, 62, 63, 70, 76, 82, 83, 84, 85, 89, 91, 92, 93, 96, 97, 98, 103, 104, 105, 106, 114, 116, 118, 120, 140
    nuclear test 19, 20, 24
Indian Ocean 20, 43
Indo-China 46, 66
Indonesia 3, 17, 18, 21, 22, 96, 98, 112 116, 118
Inspection 126
Intelligence 57 *see also* Surveillance
Iran 3, 16, 17, 18, 20, 21, 22, 111, 112, 114, 133, 136
Iraq 18, 21, 22
Ireland 74, 110
Israel 3, 17, 18, 22, 33, 59, 71, 76, 111, 112, 114, 118
    Atomic Energy Commission 2
Italy 3, 16, 21, 22, 46, 82, 85, 86, 96, 97, 100, 101, 120, 121, 124

Jamaica 96, 111, 112, 135
Japan 3, 8, 16, 17, 18, 20, 22, 26, 32, 54, 63, 67, 96, 117
Jensen, L. 96
Johnson, L.B. 100
Jordan 96, 112, 115

Kahn, H. 40
Kennedy, J.F. 65
Kenya 97, 111, 112, 113, 116, 117
Khallaf, H. (*Ambassador*) 86, 88, 102
Klein, D. (*Ambassador*) 143
Kolo, S. 106
Korea
    North 18, 21, 23, 33, 48, 59
    South 3, 18, 21, 23, 33, 48, 59, 139, 147
Kosygin, A. 100
Kosygin Proposal (1966) 44, 100, 101, 102, 103, 106, 107, 111
Kuznetsov, V.V. (*Ambassador*) 95

Laos 96
Lebanon 135
Lesotho 112
Liberia 97, 111, 135
Libya 17, 18, 21, 23, 96, 98, 111
Low Posture Doctrine *see under* Strategy for non-proliferation

MIRV (Multiple Independently Targetable Re-entry Vehicles) 135, 136
Madagascar 96, 112, 113
Maddox, J. 39, 49, 54, 56, 59, 60
Malaysia 96, 97
Malta 96, 112
Mapanza (*Ambassador*) 115
Mass destruction (weapons) 106
Mauritania 96, 98, 112, 113, 116, 117, 118
Mendlovitz, S. 12
Mexico 3, 8, 19, 21, 23, 82, 88, 89, 92, 93, 94, 103, 104, 105, 106, 107, 117, 124, 125, 134, 135, 137, 138, 141, 144
Middle East 15, 35, 60, 104, 113, 143
Mihajlović, M. (*Ambassador*) 142
'Mini-nukes' 40
Morocco 135
Mulley, F. (*Ambassador*) 92, 124
Myrdal, A. 9, 84, 93, 125

NATO (North Atlantic Treaty Organization) 16, 20, 26, 29, 38, 48, 49, 50, 53, 54, 58, 67, 75, 110
    Multilateral Nuclear Force 75
NNWS (Non-Nuclear Weapon States) *passim*
    Conference (1968) 91
    defined 2
    Group of 77 137, 144
NPT (Treaty on the Non-Proliferation of Nuclear Weapons) 1, 2, 7, 8, 11, 12, 14, 24, 27, 32, 38, 44, 49, 64, 65, 66, 72, 73, 77ff.
    draft treaty of 1965 78, 99, 118
    draft treaty of 1967 86, 105, 121
    draft treaty of 1968 91, 94, 98, 107, 108, 117, 123, 125
    future 144, 145
    negotiations (1965-68) 74ff.
    provision for review 118ff.
    Review Conference (1975) 1, 74, 121, 122, 123, 128, 129, 130ff., 144, 145, 146
NWS (Nuclear Weapon States) *passim*
    defined 1
    nuclear weapons, use of 52
Nepal 97, 111, 112, 116, 135, 142
Netherlands 3, 8, 16, 23

New Zealand 133
Nicaragua 135
Nigeria 18, 21, 23, 82, 94, 96, 102, 103, 106, 108, 124, 125, 135, 142
No-first-use *see* Agreement
Nuclear blackmail 15, 16, 17, 31, 33, 34, 44, 53, 56, 67, 69, 100, 102, 105
Nuclear capability 3, 4, 28, 39, 100
Nuclear disarmament 94, 125 *see also* Disarmament
Nuclear energy/power 2-4, 14
Nuclear explosions (peaceful) 2, 18, 30, 87, 91, 127, 130, 137
Nuclear firebreak 6, 34, 45, 59
Nuclear-free zones 42, 43, 53, 54, 56, 59, 60, 71, 100, 103, 108, 111, 127, 141, 143, 144
Nuclear fuel cycle 2, 146
Nuclear materials and facilities, export of 11
Nuclear Planning Group (NATO) 38, 50
Nuclear testing 8, 19, 24, 135 *see also* Nuclear explosions
  underground 131, 136
Nuclear weapons 13, 15, 16, 17, 19, 27, 41, 84
  delivery systems 41, 74
  deployment restrictions 42, 56, 61, 69, 80, 101, 103, 104, 142
  first-strike 6, 16, 17, 18, 111
  outlawing in space 77
  production 44, 92
  prohibition of use 103, 111
  reduction of inventories 27, 47, 69 *see also* Disarmament, arms limitation
  second-strike 5, 16, 17, 28, 40, 47, 53
  stockpiles 84
  tactical 17, 18, 28, 34, 40, 42, 53, 142, 143, 147
  treaties constraining use 52

PNE (Peaceful Nuclear Explosives) *see* Nuclear explosions (peaceful)
PTBT (Partial Test Ban Treaty) 11, 30, 52, 74, 77, 78, 79, 85, 94, 95, 119, 134
Pakistan 3, 17, 23, 58, 70, 91, 96, 97, 98, 111, 112, 114, 115, 118
Panama 96, 98, 111, 112, 117
Paraguay 110
Pariah state 17, 18, 25
Parthasarathi, G. (*Ambassador*) 114
Peaceful resolution of disputes 55, 60
Persian Gulf 35
Peru 96, 135, 141
Philippines 3, 18, 21, 23, 97, 98, 135, 140
Plutonium 2
Poland 3, 16, 23, 100
Political blackmail 108
Prestige, political 19ff., 67
  and High Posture Doctrine 36ff.
  and Low Posture Doctrine 61ff.
Proliferation, horizontal and vertical linked 27, 28, 82, 132, 133

Quester, G. 2

Ra'anan, U. 27, 28
Rhodesia (Southern) 113
Roberts, H.V. (*Ambassador*) 133
Robledo, G. 104
Robles, G. 137, 138, 144
Romania 3, 8, 16, 17, 23, 82, 86, 89, 90, 92, 93, 98, 100, 103, 106, 107, 111, 122, 123, 124, 126, 135, 136, 141, 142
Rosecrance, R. 35
Rousseau, J.J. 12
Rwanda 96, 98, 111, 112, 113, 117

SALT (Strategic Arms Limitation Talks) 63, 75, 81, 137, 142
SALT I 132
  Interim Agreement (1972) 30, 52
SALT II 41, 56, 132, 136, 138, 146
SALT III 136
Saudi Arabia 17, 18, 21, 23
Scandinavia 8
Schlesinger, J. 27, 29, 33, 35, 68
Second-strike capability *see under* Nuclear weapons
Security 7ff., 14ff., 25, 27, 31, 46, 57, 69, 72, 77, 99ff., 130ff., 138ff.
  negative guarantee 31, 42, 43, 44, 46, 49, 50, 53, 54, 56, 57, 59, 60, 61, 99, 100, 101, 102, 103, 104, 106, 108, 111, 112, 118, 127, 139, 140

positive guarantee 17, 29, 30, 31, 33, 42, 47, 48, 49, 51, 54, 56, 57, 59, 63, 66, 67, 99, 100, 101, 104, 108, 112, 116, 117, 122, 127, 140
credibility of guarantees 66ff.
Senegal 110, 135, 141
Sierra Leone 96, 98, 112, 113, 114, 116, 117
Singer, J.D. 12
Singer, M. 39, 44, 48, 49, 52, 54, 56
Smart, I. 39, 49, 50, 54, 56
South Africa 3, 18, 20, 21, 23, 59, 96, 98, 113
Spain 3, 21, 112, 118, 124, 125
Sri Lanka 98, 118 *see also* Ceylon
Status *see* Prestige
Strategic weapon reductions 53, 135, 136
Strategy for non-proliferation 25ff.
High Posture Doctrine 14, 25, 26, 27ff., 77, 87, 104, 110, 117, 138, 146, 147
Low Posture Doctrine 14, 25, 26, 27, 30, 39ff., 80, 82, 86, 88, 96, 99, 101, 104, 107, 111, 118, 121, 125, 126, 134, 137, 140, 141, 146, 147
Submarines 41
Sudan 116, 135, 141
Surveillance 75
Sweden 3, 8, 9, 16, 23, 75, 82, 84, 85, 90, 93, 94, 95, 97, 98, 103, 120, 124, 125, 130, 136, 137
Switzerland 3, 16, 23, 106, 121, 123, 125
Syria 18, 21, 23, 135, 140

TTBT (Threshold Test Ban Treaty) 75, 132, 138
Taiwan 3, 17, 21, 23
Tanzania 111, 112, 113, 115, 116, 117, 118
Testing *see* Nuclear testing
Thailand 3, 18, 21, 23, 96, 112
Third World 4, 20, 21, 22, 24
Thorsson, I. 130, 134, 137, 141, 144
Tlatelolco *see* Treaty
Treaty
ABM (1972) 30, 52, 131
ABM (1974) 30, 52
Antarctic (1959) 30, 52
Limitation of Underground Nuclear Weapons Tests (1974) 30, 52, 131
NPT *q.v.*
Outer Space (1967) 30, 52
PTBT 11, 30, 52, 74, 77, 78, 79, 85, 94, 95, 119, 134
Sea-Bed (1971) 30, 52, 131
Sino-Soviet (1950) 20
Tlatelolco 34, 43, 44, 52, 103, 111, 143
TTBT 75, 132, 138
Underground Nuclear Explosions for Peaceful Purposes (1976) 30, 52, 131
Trivedi, V.C. (*Ambassador*) 82, 89, 105
Trinidad and Tobago 96, 98, 111, 112, 116, 118
Turkey 3, 16, 18, 21, 23, 111

Uganda 96, 98, 111, 112, 113, 116, 118
United Arab Republic 3, 86, 88, 89, 90, 94, 102, 103, 104, 106, 107, 116, 124, 126
United Kingdom 2, 16, 19, 20, 62, 76, 77, 78, 79, 87, 94, 95, 100, 101, 105, 108, 109, 111, 113, 115, 118, 121, 124, 125
United Nations 7, 46, 60, 74, 82, 98, 99, 104, 110, 113, 115, 117, 129, 131, 136, 144
Charter 108, 109, 110, 112, 113, 114, 115, 117, 127, 140, 145, 147
Disarmament Commission 75, 97
General Assembly 47, 74, 77, 91, 96, 98, 103, 108, 110, 111, 112, 140
Resolution 1653 (XVI) 103, 112
Resolution 2028 (XX) 76, 82, 83, 96, 103, 117, 120, 126, 133, 141
Resolution 2153 (XXI) 104, 106, 111
Resolution 2346A, B (XXII) 91
Resolution 2372 (XXII) 98
Resolution 2373 (XXII) 116, 118
Resolution 2936 (XXVII) 131, 139, 140
Security Council 19, 63, 107, 108, 109, 110, 111, 112, 113, 114, 115, 116, 117, 118, 122, 127, 131, 138

Security Council Resolution 255
32, 33, 50, 56, 70, 107, 118, 127, 139, 140, 141, 144
USA *passim*
security guarantees 32 *see also* Security
USSR *passim*

Venezuela 21, 97, 98
Verification of agreements 8, 42, 43, 60, 81 *see also* Inspection
Viet-Nam 113
Vladivostok Accords (1974) 30, 41, 52, 131, 132, 134, 135, 137, 138

WTO (Warsaw Treaty Organization) 16, 29, 32, 48, 49, 50, 53, 54, 58, 67, 110

War
Ethiopia (1936) 46
Indo-China 46, 66
Middle East (1973) 4
World War II 19
Western European Community 16, 20, 23, 75
World Disarmament Conference 131
Wright, M. 12

Yugoslavia 3, 8, 16, 17, 23, 96, 97, 98, 111, 112, 135, 140, 141, 142

Zaire 18, 21, 23, 135, 141
Zambia 96, 98, 112, 113, 114, 115, 116, 118
Zollner, M.-L. (*Ambassador*) 113, 117